Sustained Simulation Performance 2013

Michael M. Resch • Wolfgang Bez • Erich Focht
Hiroaki Kobayashi • Yevgeniya Kovalenko
Editors

Sustained Simulation Performance 2013

Proceedings of the joint Workshop on Sustained Simulation Performance, University of Stuttgart (HLRS) and Tohoku University, 2013

Springer

Editors
Michael M. Resch
Yevgeniya Kovalenko
High Performance Computing Center
Stuttgart (HLRS)
University of Stuttgart
Stuttgart
Germany

Wolfgang Bez
NEC High Performance Computing
Europe GmbH
Düsseldorf
Germany

Erich Focht
NEC High Performance Computing
Europe GmbH
Stuttgart
Germany

Hiroaki Kobayashi
Cyberscience Center
Tohoku University
Sendai
Japan

Front cover figure: Volume renderings of a human Femur, a femoral head and cubes of cancellous bone micro structure with 1.2mm edge length. The cubes are used as base data to derive continuum mechanical material properties via direct numerical simulation. Illustration by Ralf Schneider, High Performance Computing Center Stuttgart, Stuttgart, Germany

ISBN 978-3-319-01438-8 ISBN 978-3-319-01439-5 (eBook)
DOI 10.1007/978-3-319-01439-5
Springer Cham Heidelberg New York Dordrecht London

Library of Congress Control Number: 2013947943

Math. Subj. Class. (2010): 68Wxx, 68W10, 68Mxx, 68U20, 76-XX, 86A10, 70FXX, 92Cxx, 65-XX

Printed on acid-free paper

Springer is part of Springer Science+Business Media (www.springer.com)

Preface

The field of high-performance computing is currently witnessing a significant shift of paradigm. Ever larger raw number crunching capabilities of modern processors are in principle available to computational scientist. However, efficiently exploiting modern processors is getting more and more complex; this is particularly true for so-called accelerators as GPGPUs or many-cores as the MIC architecture. (GPGPU, many-cores, and MIC are very common technical terms which are not introduced in the preface, but in the later chapters).

On the other hand, many areas of computational science have reached a saturation in terms of problem size. Scientists often do no longer wish to solve larger problems. Instead they wish to solve smaller problems in a shorter time. The current architectures, however, are much more efficient for large problems than they are for the more relevant smaller problems.

This series of workshops focuses on *sustained simulation performance*, i.e., high-performance computing for real application use cases, rather than on peak performance, which is the scope of artificial benchmarks. The series was established in 2004, initially named Teraflop Workshop, and renamed to Workshop on Sustained Simulation Performance in 2012. In general terms, the scope of the workshop series has focussed on issues related to achieving high sustained performance on all kinds of architectures – from vector systems, to accelerator-based architectures, to clusters and state-of-the-art MPP systems. Special emphasis is given to programmer productivity in the setting of current and future trends in hardware and software developments.

This book presents the results of the 16th and 17th installment of the series. The 16th workshop was held at the High-Performance Computing Center, Stuttgart, Germany, in December 2012. The 17th workshop was held in March 2013 at the headquarters of NEC Corporation in Tokyo, Japan, and organized jointly with the University of Tohoku, Sendai, Japan.

The topics studied by the contributed papers include analysis and extrapolations of performance and energy efficiency of modern hardware architectures (Part I), frameworks and libraries aiming at increasing programmer productivity on future systems (Part II), and application use-cases studies (Part III).

We would like to thank all the contributors of this book and the Sustained Simulation Performance project. We thank especially Prof. Hiroaki Kobayashi for the close collaboration over the past years and are looking forward to intensify our cooperation in the future.

Stuttgart, Germany
July 2013

José Gracia
Michael Resch

Contents

Part III Computational Engineering Applications and Multi-Physics Simulations

Part I
Challenges of Modern HPC Systems: Performance and Energy Efficiency Analysis

Feasibility Study of Future HPC Systems for Memory-Intensive Applications

Hiroaki Kobayashi

Abstract After the successful launch of K-Computer in Japan, the Japanese government started a new R&D program entitled "Feasibility Study of Future HPCI Systems." In this program, social and scientific demands for HPC in the next 5–10 years will be addressed, and HPC systems that satisfy the demands and key technologies to develop such systems will be discussed and evaluated. Currently, three system design teams get involved in this program, and this article present a HPC project entitled "Feasibility Study of Future HPC Systems for Memory Intensive Applications," which is conducted by a team of Tohoku University, JAMSTEC and NEC.

1 Introduction

Since the first peta-flop/s machine named Roadrunner became the world's first TOP500 LINPACK system in 2008, about 30 peta-flop/s systems have been installed around world; US, Germany, Italy, France, UK, Australia, Russia, China and Japan only within 4 years. Now the hot topic in the HPC community is when and where the first exascale system will become available. Although *exascale* does not exactly mean exa-flop/s, when taking a look at the trend in sustained LINPACK performance in TOP500, it takes 12 years from 1 tera-flop/s machine named ASCI RED developed in US in 1996 to 1 peta-flop/s Roadrunner in 2008 [7]. US, Europe, China and Japan started several HPC strategic programs for targeting at realization of exascale systems around 2020.

In Japan, after the successful launch of K-Computer, which was the first 10-peta flop/s LINPACK system in 2011, MEXT (Ministry of Education, Culture,

H. Kobayashi (✉)
Tohoku University, Sendai 980-8578, Japan
e-mail: koba@isc.tohoku.ac.jp

M.M. Resch et al. (eds.), *Sustained Simulation Performance 2013*,
DOI 10.1007/978-3-319-01439-5_1,

Sports, Science and Technology) organized a committee to discuss the HPC policy of Japan for the next 5- to 10-year research and development on national leading-supercomputers, and the committee decided to start a program entitled *Feasibility Study of Future HPCI systems* last year. The objectives of this program is to

- Discuss future high-end systems capable of satisfying the social and scientific demands for HPC in the next 5–10 years in Japan, and
- Investigate hardware and software technologies for developing future high-end systems available around year 2018 that satisfy the social and scientific demands.

After the review and selection of the proposals to this program, three teams, which are University of Tokyo with Fujitsu (Project Leader: Professor Yutaka Ishikawa), University of Tsukuba with Hitachi (Project Leader: Professor Mitsuhisa Sato), and Tohoku University with NEC (Project Leader: Hiroaki Kobayashi), started the feasibility studies as a 2-year national project.

In this article, we present an overview of our project entitled, "*Feasibility Study of Future HPC Systems for Memory-Intensive Applications*." Section 2 describes our system design concept with discussing target applications that would be solutions to several important issues as social and scientific demands around 2020. In addition, the basic specification and configuration of the system are presented. Section 3 summarizes the current state of the project and its future plan.

2 Design Concept of the Target System

In the last decades, microprocessors for high-end computing systems boost their flop/s rates by introducing multi- and many-core architectures into the chip design. However, their off-chip memory throughputs are not improved well, and as a result, the bytes per flop rate, B/F, which is a ratio of the memory bandwidth to the peak flop/s is decreasing as shown in Fig. 1 [6]. One exceptional case that keeps B/F high is the vector processor developed by NEC for its SX supercomputer series. The latest system of the NEC supercomputer is SX-9, and its processor provides a 102.4 Gflop/s as a single-core processor with the 256 GB/s memory interface, resulting in the significant B/F rate of 2.5, compared to 0.5 or lower B/F rates of modern microprocessors such as IBM power7 [1], Intel Xeon [3] and Fujitsu SPARC IVfx [4]. However, the vector processor is also facing the memory wall problem as the vector processor flop/s is also increasing.

The memory bandwidth is a key factor to exploit system peak performance and make simulation much more efficient and productive runs. As the B/F is decreasing, it becomes more difficult to feed the necessary data to plenty of arithmetic units on a chip, resulting in lowering the processing efficiency. Figure 2 shows the attainable performance of applications on several modern microprocessors as a function of application B/F rates. An application B/F rate is the memory access intensity of an application, and shows the amount of memory access in bytes per one floating operation in its highest cost kernel. In the figure, "*Roof Lines*" of processors are depicted.

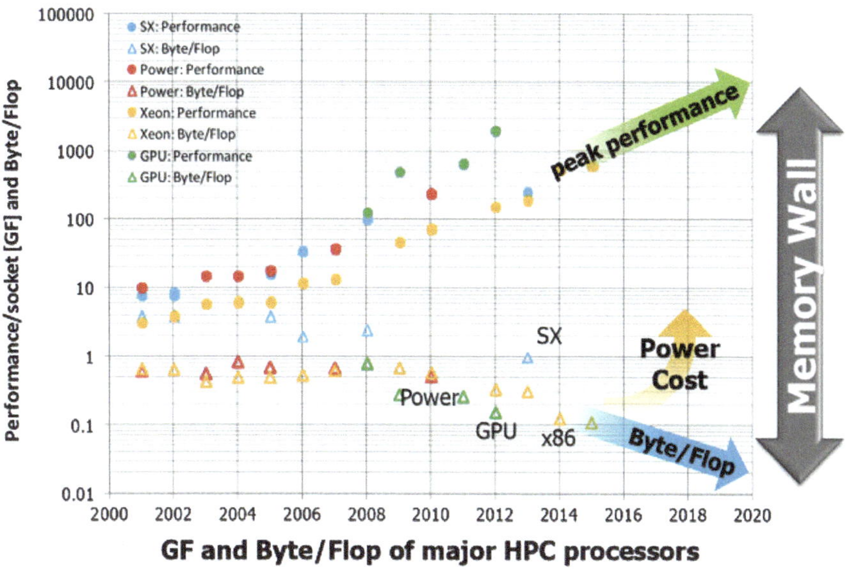

Fig. 1 Trend in processor peak flop/s and B/F

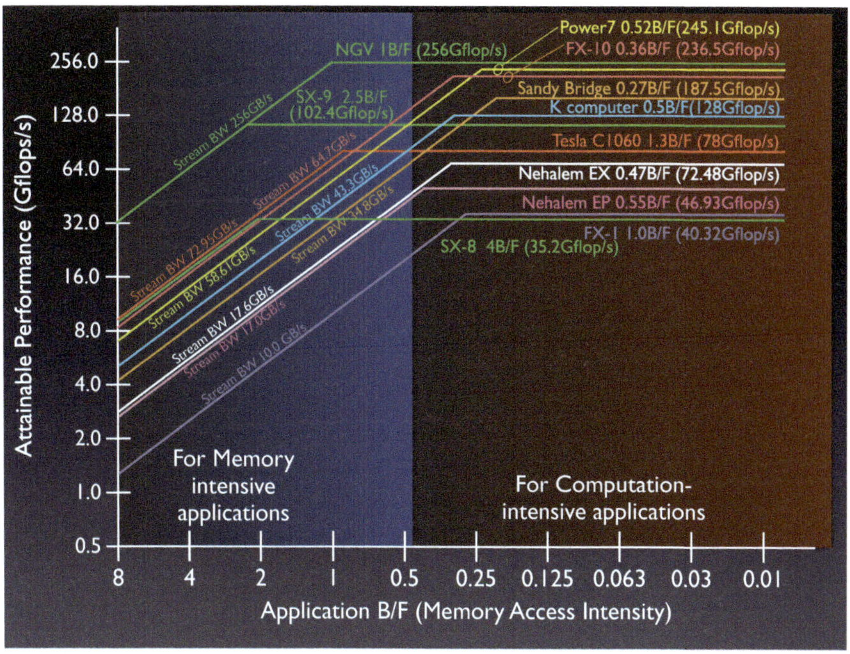

Fig. 2 Roofline models of modern microprocessors

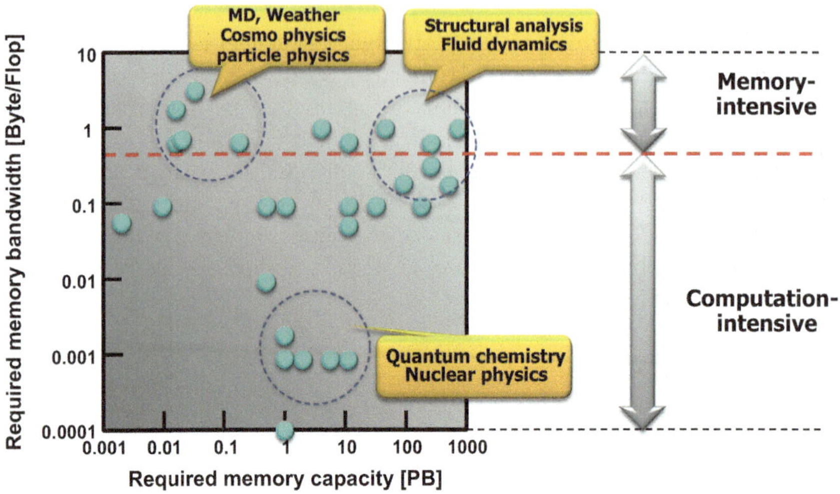

Fig. 3 Memory requirements of applications

A flat roof line shows the upper limit of attainable performance of an application defined by the peak performance of a processor on which the application runs, and a slant roof line is given by the limitation of memory bandwidth of the processor when the application B/F is larger than the processor B/F. As the figure suggests, the order of the processors regarding attainable performance in the case of memory-intensive applications is quite different from that in the computation-intensive applications. For memory-intensive applications, the attainable performance is far from peak-performance of processors when their B/F rates do not match the application B/F rates.

Unlike LINPACK, sustained performance of many practical applications is memory-limited, because these applications need a lot of data during operations in their high cost kernels. According to the application development roadmap report summarized in Japan in 2012 [5], many important applications in the wide variety of science and engineering fields need 0.5 B/F or more, as shown in Fig. 3. Therefore, if we continue to develop high-end computing systems by concentrating on increasing flop/s rates, simply targeting toward exa-flop/s in 2020, rather than memory bandwidth, their applicable area are getting limited, i.e., there will be a high probability that only few % of peak performance of exa-flop/s would be effective in the execution of practical exascale applications, and lots of arithmetic unites end up wasted during their execution.

Based on the above observation, in our project, we are much more focusing on device and architectural technologies to keep high B/F in the design of future HPC systems for exascale computing of important applications available around 2020. In particular, we will explore the design space of the future HPC systems to make them more flop/s-efficient and applicable to the wide fields of science and

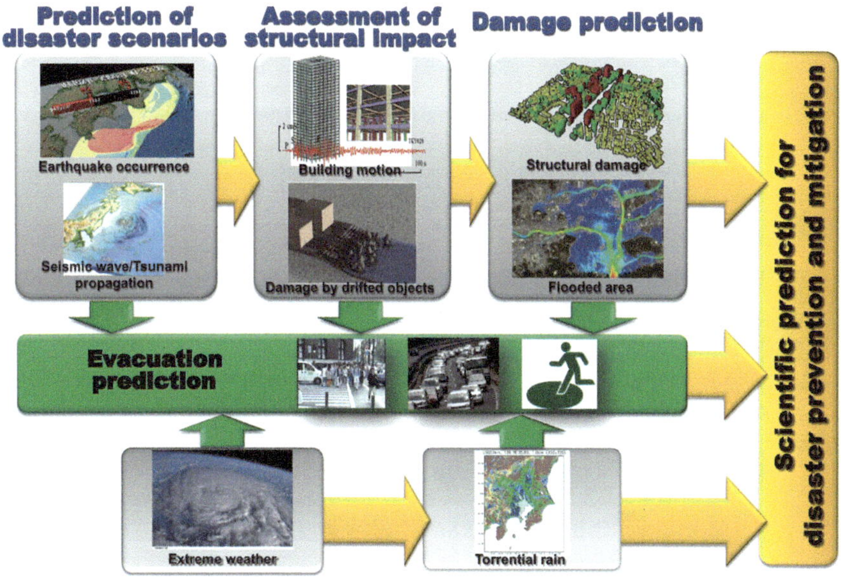

Fig. 4 Simulation model of compound disaster

engineering, especially satisfying social demands for realizing the safe and comfortable society with the HPC technology. To this end, we bring many leading application researchers into the project for co-design of architectures and applications, in order to introduce applications viewpoints into the system design, and make the system highly optimized and friendly in both performance per power/energy consumption and programming, respectively.

Although many applications need higher B/F rates in their programs, important memory-intensive applications come from earth sciences and engineering fields. Especially, after the great East-Japan earthquake in 2011, there is a growing demand for the prevention and mitigation of natural disaster by large earthquakes and tsunami in Japan. Therefore, we study the simulation for dealing with compound disaster, which consists of an underground structure model, an strong motion model, an tsunami model, and a whole city model, in order to provide useful information about damage of lands and building, and evacuation guidance after large earthquakes and tsunami as shown in Fig. 4. In this compound disaster simulation model, the weather model that simulates typhoon, concentrated heavy rains, and tornado is also combined as another type of disaster sources that needs to be considered.

In addition to natural disaster analysis as memory-intensive applications, we also address engineering simulations as our memory-intensive applications, in particular, high-resolution airplane models for digital flight and multi-physics models for reliable and efficient turbo machineries for power plant systems, which need exascale computing as shown in Fig. 5. For these target applications, we investigate their fundamental simulation models available in 2020, and estimate B/F of their

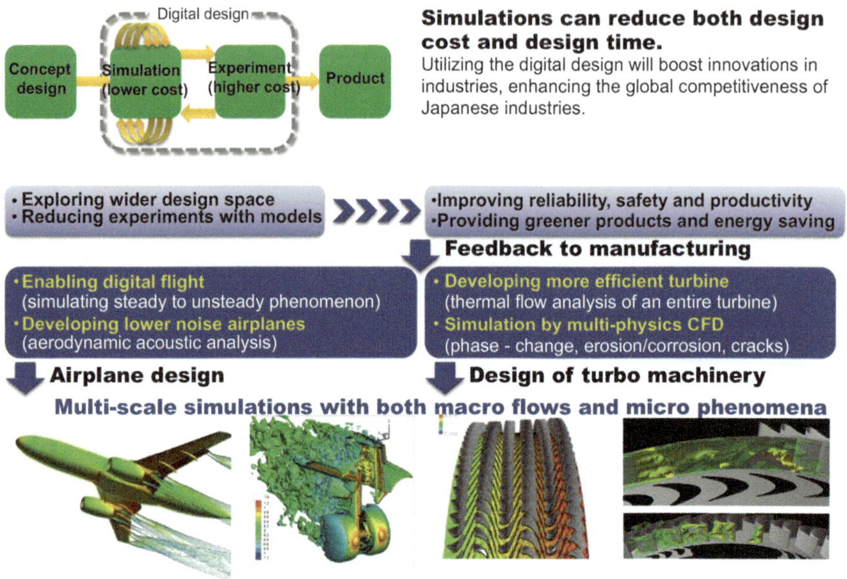

Fig. 5 Advanced digital design of airplanes and turbo machineries

high cost kernels, the total amount of the required computation, memory capacity, and the time limitation to the solution, for the productivity of simulation as basic requirements when considering the design of the system architecture for future HPC systems whose design target is around 2018. Table 1 summarizes requirements of the applications. In the top of the table, applications examined in this project are presented, and the simulation flow in the dotted box shows compound disaster analysis. In the bottom of the table, there are two categories for applications: one is for applications that need very high-resolution models for their single run. The other is for applications each of which is a collection of several runs of moderate resolution models with different simulation parameters. The table suggests that all the applications for disaster analysis and engineering product design need two or more B/F rates in their high cost kernels, and one to hundreds exascale floating point operations should be processed less than 10 h down to in 30 min for their simulation productivity. They also need up to 14 PB for memory space.

After the careful review of these application requirements for exascale computing, we reached the following basic design concepts:

- Design an advanced vector multi-core architecture optimized not only for long vectors but also for short vectors with indirect memory access,
- Design a larger computing node connected to larger shared memory at higher B/F rate, compared to conventional microprocessors and their successors projected, in order to keep the number of parallel processing nodes required by target applications as low as possible,

Table 1 Requirements of target applications for the design of the HPC system

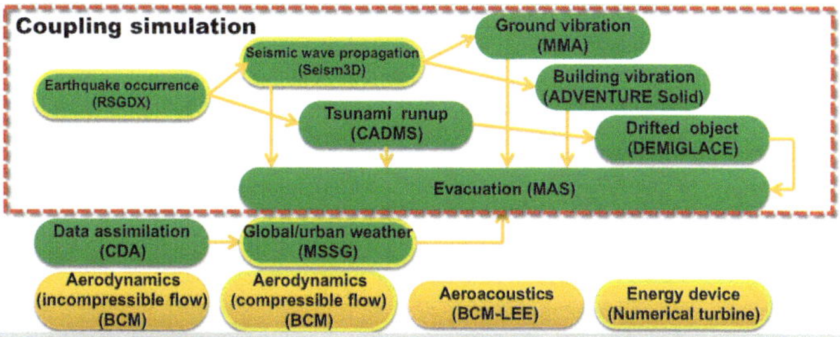

Purpose	Application	B/F	# of floating point operations ($\times 10^{18}$)	required memory size	Max. allowed runtime (hr)
Super-high resolution single run	RSGDX	4.50	230	1.5 PB	10
	Seism3D	2.14	160	9.6 PB	2
	MSSG	4.00	720	175 TB	6
	BCM	5.47	1	13.6 TB	0.5
High-resolution ensemble run	Comprehensive disaster prevention	2~5	100	4 PB	2~3
	Numerical turbine	2.33	140	163.5 TB	20

- Design a advanced memory subsystem using innovative 2.5D silicon interposer and 3D-die stacking technologies to satisfy the requirements of B/F rates of 2 or more of target applications,
- Design a network system with better local communications, while keeping the latency of global communication as low as possible,
- Design a hierarchical storage and I/O subsystem that satisfies the requirements for data assimilation in disaster analysis, and
- Design a system software that provides the standard programming environment, i.e., LINUX-based system software, but it should be customized for exploiting the potential of the advanced vector architecture without sacrificing the standard programming environment.

Figure 6 shows a block diagram of the system designed based on these concepts. The basic specification of the system is as follows:

- Performance/Socket
  - >1 Tflop/s of a peak performance
  - >1 TB/s of a memory bandwidth
  - >128 GB of memory capacity
- Performance/Node
  - Up to 4 sockets/node
  - >4 Tflop/s of a node peak performance
  - Up to 1 TB of the shared memory

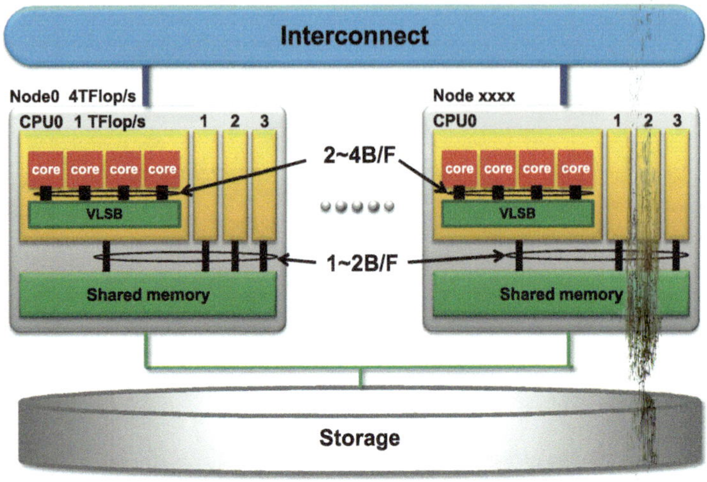

Fig. 6 Conceptual design of the target system

The system scale will be decided so as to satisfy the requirements regarding *the time to solution* of individual applications after the performance estimation.

The design concepts presented here are very aggressive and challenging, and we understand that there several obstacles that should be solved before going into the production of the above system mainly due to cost and power budget in addition to technological issues, but application designers, architecture designers and device designer in our project work tightly together to make innovation happen for realizing highly productive HPC systems that will be expected to became available around 2018!

3 Summary

This article describes our on-going project for design and development of future HPC systems, especially suited for memory-intensive applications. Our design is based on the vector architecture, but many advanced technologies for efficient processing of short vectors with indirect memory references will be introduced. New device technologies such as 2.5D interposer and 3D die-stacking for the design of the memory subsystem, which make the memory subsystem high-bandwidth and low-power, will aggressively be applied to the system design. For the true holistic collaboration among applications, architectures and device technologies, we bring leading researchers and engineers in individual fields nation-wide into the project. Now we are finalizing the details of the architecture and estimating its performance by using target applications scaled to exa-level computing, which satisfy the social demands and make science and engineering breakthrough happen.

Our design philosophy is to *waking up plenty of sleeping floating point units on a chip by improving memory performance for practical applications, not optimized for LINPACK.* We believe the brute force to exascale computing by simply increasing the number of simple cores on a chip does not make sense any more for practical applications in the wide variety of science and engineering. So far, LINPACK of TOP 500 drives the development of high end systems as their performance measure, however, many people are now aware of the limitation of LINPACK for the evaluation of productivity of high-end computing systems [2] We think that innovative memory system design would be a key to realization of highly productive HPC systems in the next 5–10 years, and we believe that our approach will open up the new way to exascale computing. Let' make the supercomputer for the rest of us happen, not for LINPACK only!

Acknowledgements Many colleagues get involved in this project, and great thanks go to Dr. Y. Kaneda and Dr. K. Watanabe of JAMSTEC (Japan Agency for Marine-Earth Science and Technology) as co-leaders of the application group, Professor M. Koyanagi of Tohoku University as the leader of the 2.5D/3D device group, and Ms. Y. Hashimoto of NEC as the leader of the NEC application, system and device design group. This project is supported by MEXT.

References

1. M. Floyd et al. Harnessing the Adaptive Energy Management Features of the POWER7 chip. In *HOT Chips 2010*, 2010.
2. M. Flynn et al. Moving from petaflops to petadata. *Communications of the ACM*, 56:39–42, 2013.
3. Oded Lempel. 2nd Generation Intel* Core* Processor Family: Intel Core i7, i5 and i3. In *HOT Chips 2011*, 2011.
4. Takumi Maruyama. SPARC64(TM) Viiifx: Fujitsu's New Generation Octo Core Processor for PETA Scale Computing. In *HOT Chips 2009*, 2009.
5. MEXT HPC Task Force. (in Japanese) Report on Application R&D Roadmap for Exascale Computing. 2012.
6. S. Momose. Next Generation Vector Supercomputer for Providing Higher Sustained Performance. In *COOL Chips 2013*, 2013.
7. Top 500 Supercomputer Sites. http://www.spec.org/.

Analysing the Performance Improvements of Optimizations on Modern HPC Systems

Kazuhiko Komatsu, Toshihide Sasaki, Ryusuke Egawa, Hiroyuki Takizawa, and Hiroaki Kobayashi

Abstract Recently, there are many types of supercomputing systems being equipped with vector processors, scalar processors, and accelerators as processing elements of the systems. Although all kinds of calculations cannot effectively be performed on one HPC system, a part of calculations can exploit the potential of a HPC system by considering both the calculations and the system. These tendencies that each HPC system is designed and suitable for specific fields of calculations continue in order to achieve higher performance for target HPC codes. Therefore, even though the same HPC code is executed on multiple HPC systems, the sustained performances on HPC systems are different. As characteristics of a HPC code mainly depend on optimization methods, clarifying the performances by the optimization methods on multiple HPC systems becomes important for developing performance-portable HPC codes, which can exploit the potential of every HPC system. By considering both the optimization methods and the HPC systems, this paper clarifies the performances of the optimization methods on multiple HPC systems.

K. Komatsu (✉) · R. Egawa · H. Kobayashi
Cyberscience Center, Tohoku University/JST CREST, 6-3 Aramaki-aza-aoba, Aoba, Sendai 980-8578, Japan
e-mail: komatsu@isc.tohoku.ac.jp; egawa@isc.tohoku.ac.jp; koba@isc.tohoku.ac.jp

T. Sasaki
Department of Information and Intelligent Systems, Tohoku University, 6-6-01 Aramaki-aza-aoba, Aoba, Sendai 980-8579, Japan
e-mail: toshihide@sc.isc.tohoku.ac.jp

H. Takizawa
Graduate School of Information Sciences, Tohoku University/JST CREST, 6-6-01 Aramaki-aza-aoba, Aoba, Sendai 980-8579, Japan
e-mail: tacky@isc.tohoku.ac.jp

M.M. Resch et al. (eds.), *Sustained Simulation Performance 2013*,
DOI 10.1007/978-3-319-01439-5_2,

1 Introduction

Due to the rapid advancements in semiconductor technologies, the number of transistors that can be integrated on a processor is drastically increased. As a large number of transistors has brought the design space of a processing element wider, various processor architectures have been developed. For example, recent scalar processors are equipped with multiple high performance cores and cache memories with large capacity . In addition, some scalar processors are equipped with graphics processing cores.

As there are many types of processors, the types of the HPC systems that support large scale HPC applications have also been increasing [1]. Vector supercomputing systems employ vector processors and can calculate sets of data elements at the same time by utilizing its wide sustained memory bandwidth [2, 6]. Scalar supercomputing systems generally consist of a large number of scalar processors and can perform parallel calculations by utilizing the massively parallelism of an application [3]. Accelerator type supercomputing systems have both host processors and accelerators such as GPUs. It can effectively perform data parallel applications by exploiting high floating point operations and memory bandwidth of the accelerators. [4, 5].

As each HPC system has its own advantages and disadvantages, the sustained performances of the same HPC code executed on various HPC systems are different. If a HPC code is suitable for one HPC system, it can efficiently exploit the potential of that HPC system. However, if the HPC code is not suitable for the characteristics of another HPC system, the potential of the HPC system is not effectively utilized. This is because the characteristics of a HPC code are not always suitable for every HPC system.

As the characteristics of HPC codes change by optimizations applied to the HPC codes, the performances of the optimizations on multiple HPC systems should be clarified for understanding the characteristics of HPC codes. By considering the performances of the optimizations among multiple HPC systems, performance-portable HPC codes that exploit the potential of various HPC systems can be developed.

This paper firstly shows performance differences of the optimization methods among various HPC systems. Then, to clarify the features and performances of the optimizations among various HPC systems, the performance of each optimization method that is widely utilized for many HPC codes is evaluated. By considering both optimization methods and HPC systems, this paper discusses the performances of the optimization methods on multiple modern HPC systems.

The rest of this paper is organized as follows. Section 2 shows the performance differences among multiple platforms through preliminary evaluations of practical HPC applications. Section 3 briefly describes optimization methods widely applied to practical HPC applications on multiple HPC systems. Section 4 clarifies the performances of the optimization methods on various HPC systems through the evaluations by considering both the optimization methods and HPC systems. Finally, Sect. 5 gives conclusions of this paper.

2 Performance Differences Among Multiple Modern HPC Systems

In this section, performance differences among multiple HPC systems are discussed using kernels excerpted from practical HPC codes. By using multiple HPC systems that include a vector supercomputing system and several scalar supercomputing systems, the preliminary performance evaluations of the practical HPC codes are conducted. From these results, the performance differences among multiple HPC systems are shown.

Figure 1 shows the performance improvements of each HPC system by removing undefined variables from an original kernel of a practical HPC code. The x-axis indicates the HPC systems used for the preliminary evaluations. The y-axis indicates the speedup ratio of the optimized code to the original code on each HPC system. In the original kernel, variables in a loop are assigned only when a branch condition is true. Otherwise, the variables are not undefined, and then unused. In such a situation, a compiler cannot generally know whether there is a dependency among iterations in the loop. By initializing such undefined variables using the same branch condition before the loop, a compiler can know the dependency among the iterations and may apply further automatic optimizations.

In Fig. 1, the performance improvements by removing undefined variables can be observed in Hitachi SR16000 M1 and NEC SX-9. The performances of the optimized kernel on the other systems are almost the same as the original kernel. The reason of the same performances is that the compilers of both Fujitsu FX1 and Fujitsu FX10 might successfully detect the independence among the iterations. Thus, the same performances between the original kernel and the optimized kernel can be achieved. Generally, as variables should be initialized, the performances by removing undefined variables are the same or improved.

Figure 2 shows performance improvements by combinations of multiple optimization methods such as a loop collapsing, a loop exchange, an invariant-code motion, and uses of mask operations. These multiple optimization methods are applied to a kernel of another practical HPC code. Different form Fig. 1, only NEC SX-9 achieves about 13 times performance improvements. However, the sustained performances of the other HPC systems drastically degrade by the optimizations. Only 10–60 % of the performances compared with the original kernel is obtained. Although the original kernel and the optimized kernel running on the HPC systems are the same, the sustained performances are completely different. This is because the optimized kernel are not always suitable for every HPC systems. By decomposing the combinations of the optimization methods into each optimization method and analysing the performance improvements, the optimization methods should be clarified.

In order to clarify the performances of the optimization methods on the performance among multiple HPC systems, the widely used optimization methods for HPC codes are selected and evaluated in this paper.

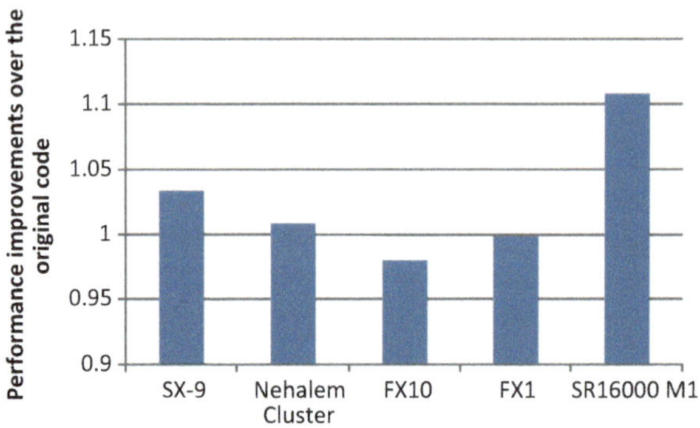

Fig. 1 Speedup of each HPC system by removing undefined variables

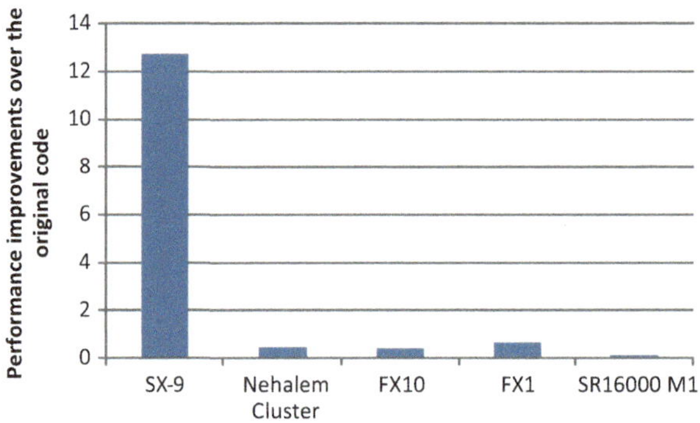

Fig. 2 Speedup of each HPC system by multiple optimization methods

3 Optimization Methods Widely Used in HPC Codes

In this section, the widely used optimization methods in HPC codes are briefly described.

3.1 Uses of Temporal Variables

When the same data are utilized multiple times in the same iteration of a loop, the load instructions of the same data might be performed multiple times. Storing such reusable data into temporal variables can avoid redundant memory instructions.

However, as the limited registers in a processor are used for temporal variables, register spilling that some variables are transferred from registers to memory may occur due to the shortage of the registers. As register spilling causes more memory accesses, the abuse of temporal variables may cause performance degradation.

3.2 Loop Distribution

A loop distribution divides a loop into multiple loops with the same loop index. A loop body is divided into multiple parts of the loop body. A loop distribution prevents register spilling that easily occurs when a lot of calculations is necessary in one loop. Furthermore, by dividing a loop body, a loop distribution enhances the locality of data required in a part of the loop body. Storing the data with high locality into on-chip memory brings effective uses of on-chip memory.

On the other hand, loop overhead such as checking loopa conditions and loop controls might cause performance degradation since the number of loops increases by a loop distribution.

3.3 Loop Unswitching

A loop unswitching moves a conditional branch from inside a loop to outside the loop. Since a loop unswitching can remove the conditional branch in a loop, a loop body might be further optimized by a compiler and executed efficiently. However, as the same loop body is described several times after the conditional branch, the maintainability of the code decreases.

3.4 Loop-Invariant Code Motion

In order to remove the same calculations among iterations, a loop-invariant code motion moves loop-independent calculations to outside of a loop. As a loop-independent code motion can remove the redundant calculations in a loop, the performance improvement can be expected.

3.5 Loop Collapsing

A loop collapsing makes the length of a loop long by collapsing triple nested loops into a single nested loop for effective vector and SIMD processing. If the length of a loop is long enough, the more elements can be calculated at the same time by vector and SIMD processing.

However, the number of memory accesses in a loop may increase as well as loop unrolling. It causes low utilization of on-chip memory due to the reduction in the temporal locality in a loop.

3.6 *Loop Exchange*

A loop exchange changes an inner loops with an outer loop. As a loop exchange changes the order to access data, the locality of reference might be improved, resulting in efficient memory accesses and uses of on-chip memory.

3.7 *Uses of Mask Operations*

Conditions of all branches are calculated in advance and the results of the conditions are stored in a mask array. According to the results of the condition in a loop body, appropriate calculations are decided. Thus, a compiler tries to perform such a branch condition using vector and SIMD operations with mask operations. As a result, as conditional branches are performed by vector and SIMD processing, the performance improvement can be expected. However, unless such vector and SIMD processing is performed, the conditional branch in a loop body may drastically degrade the performance of the calculations.

3.8 *Loop Unrolling*

Loop unrolling duplicates a loop body with a different loop index, which can reduce the amount of loop overhead such as checking loop conditions and loop controls. Moreover, by unrolling the loop body, the number of memory accesses to the same data used in different iterations can be reduced. In addition, by unrolling the short length of loops, the automatic vectorization and SIMDization by a compiler can be enhanced using the outer side of a loop body.

On the other hand, as the temporal locality in a loop is reduced, the utilization of the on-chip memory might become low.

4 Performances of Optimizations on Multiple HPC Systems

4.1 *Experimental Environments*

From three scientific and engineering applications running on Cyberscience Center, Tohoku University, three kernels are excerpted from their dominant calculations, each of which has two versions. One is the original code written in the practical applications. The other is the optimized code that is applied one or more optimization methods described in Sect. 3.

Table 1 indicates the four kernels and the applied optimization methods to the optimized version of the kernels, whose optimizations are reviewed in Sect. 3.

Table 1 Four kernels of the practical applications

# of Kernels	Optimization method
Kernel1	Uses of temporal variables
Kernel2	Loop distribution
Kernel3	Loop unswitching

Table 2 Node Specification of HPC systems used in the evaluations

HPC System	Peak Gflops/s	Sockets per node	Cores per socket socket	Memory BW GB/s	On-chip memory	B/F
NEC SX-9	1676.8	16	1	256	256 KB ADB	2.5
Intel Nehalem EX	289.92	4	8	34.1	256 KB L2/core, 24 MB shared L3	0.47
Fujitsu FX1	41.28	1	4	40	6 MB shared L2	1.0
Fujitsu FX10	236	1	16	85	12 MB shared L2	0.36
Hitachi SR16000 M1	980.48	4	8	128	256 KB L2/core, 32 MB shared L3	0.52

The five HPC systems, whose specifications are shown in Table 2, are utilized for the evaluations. These HPC systems are classified into a vector system and scalar systems. NEC SX-9 is a vector parallel supercomputer consisting of a large symmetric multi-processing (SMP) nodes, each of which has 16 102.4 Gflop/s-vector processors. The effective uses of an on-chip 256 KB cache named Assignable Data Buffer (ADB) in NEC SX-9 are a key to exploit the potential of NEC SX-9. As the main memory and ADB can simultaneously provide data to vector pipelines, the vector processor can access those data at a high sustained bandwidth.

Intel Nehalem EX cluster, Fujitsu FX1, Fujitsu FX10, and Hitachi SR16000 M1 are scalar parallel supercomputers that are equipped with Nehalem EX, SPARC64VII, SPARC64IXfx, and Power7 processors, respectively. As shown in Table 2, these scalar processors also have large on-chip cache memories. On-chip L2 and/or L3 caches should be used for data with high locality to shorten memory access latency. Moreover, uses of SIMD instructions are essential to efficiently process multiple data.

4.2 *Performance Improvement of Optimizations among Multiple HPC Systems*

In order to examine the differences of performance improvements by the optimization methods on the multiple HPC systems, the speedup ratio of the optimized kernel to the original kernel on each HPC system are utilized. The speedup ratio is shown as following equation.

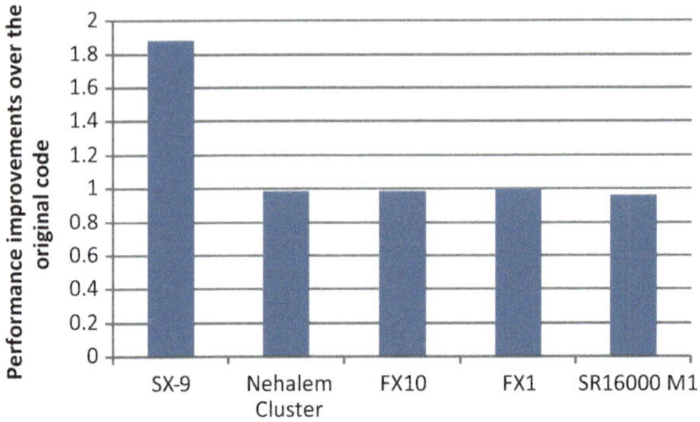

Fig. 3 Speedup of each HPC system by uses of temporal variables

$$Speedup_{System} = \frac{Time_{System}^{Original}}{Time_{System}^{Optimized}}, \tag{1}$$

where $Speedup_{System}$ indicates the speedup ratio of the optimized kernel on a HPC system. $Time_{System}^{Optimized}$ and $Time_{System}^{Original}$ indicate the execution times of the optimized and original kernels, respectively. By comparing the speedup ratios among multiple HPC systems, the performance differences can be examined.

Figures 2 and 3 show that the comparisons of the speedup ratios on multiple HPC systems. The x-axis indicates the HPC systems and the y-axis indicates the speedup ratios of the optimized kernel to the original kernel on all cores in each system.

Figure 3 shows the speedup ratios of kernel1. In the optimized kernel, a variable that is loaded from the same array twice in one iteration is stored into a temporal variable, which may reduce of the number of load instructions. Figure 3 shows that only NEC SX-9 achieves about 1.9 times performance improvements. As the memory access latency of NEC SX-9 is larger than those of scalar supercomputing systems, the second load instruction is issued before the first load instruction is complete. The use of a temporal variable avoids two load instructions for the same variable, which could remove the second load instruction. However, the use of a temporal variable does not contribute to the performance improvements on the scalar supercomputing systems. This is because the reusable variable is automatically stored in the large capacity of the cache memory in the scalar processors. Thus, the explicit use of temporal variables is not necessary for the scalar processors. As a result, uses of temporal variables are useful for a HPC system whose processor needs long memory latency when multiple load instructions to the same data issue in a short period time.

Figure 4 shows the speedup ratios of kernel2. As the large loop body in the original kernel may cause register spilling, loop distributions are applied to the optimized kernel. Figure 4 shows that NEC SX-9 achieves about 2.4 times

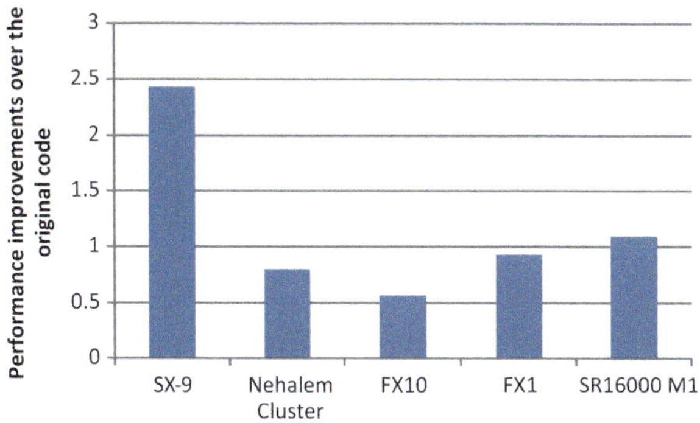

Fig. 4 Speedup of each HPC system by loop distributions

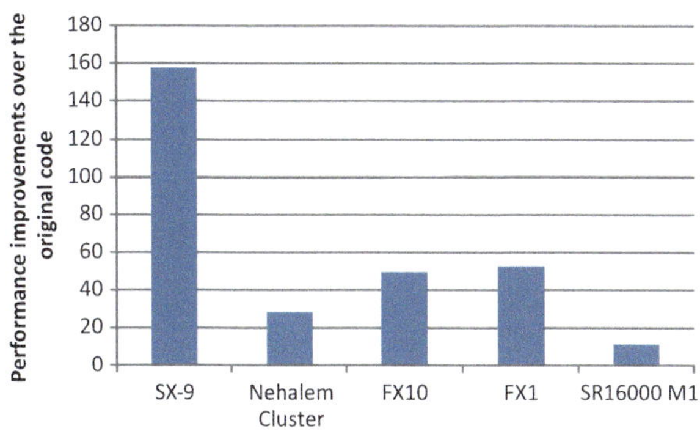

Fig. 5 Speedup of each HPC system by a loop unswitching

performance improvements because the enough number of registers in one iteration of a loop can be utilized. Furthermore, effective uses of ADB in which more reusable data can be stored further improve the performance. For Hitachi SR16000 M1, which has more registers than the other scalar systems, the loop distributions are effective. Hitachi SR16000 M1 achieves about 10 % performance improvements. The sustained performances of the other scalar systems, however, degrade due to the shortage of the registers and increase of the conditions of the loop. The divided loop bodies are still large for the scalar processors that have fewer registers than NEC SX-9 and Hitachi SR16000 M1. As a result, loop distributions considering the appropriate number of registers required for a kernel on each HPC system is necessary.

Figure 5 shows the performances of the loop unswitching for Kernel 3 among multiple HPC systems. As a variable used for a branch condition is unchanged inside a loop, the branch condition can be moved to outside of the loop. Figure 5 shows that the loop unswitching achieves about 11–158 times performance improvements and it is very effective for all HPC systems. For the vector supercomputing system, removing the branch calculations inside a loop by the loop unswitching is particularly effective because vector processors are not essentially good at branch calculations. Since branch conditions in a loop body basically reduce the sustained performance for every HPC system, a loop unswitching should positively apply if branches can move to outside of a loop.

4.3 Analysis of Combination of Optimizations

Analysing the optimized kernel that is applied the several optimizations, which is shown in Sect. 2, the effects of the multiple optimizations are further clarified. In the optimized kernel, a loop collapsing, a loop exchange, an invariant-code motion, and uses of mask operations are applied. To investigate the performance of each optimization and its combinations of these optimizations, the optimized kernel with the multiple optimizations is decomposed into several kernels, each of which applied only one optimization or their combinations.

Figure 6 shows the normalized execution times of each optimization and their combinations. Only one core is used for the evaluation to simplify the experimental environment. Furthermore, no automatic optimization is applied by a compiler. Thus, the performance of only the optimization can be examined. From this figure, most of each optimization and its combinations are effective for all HPC systems. However, in the case of trying to use mask operations for efficient vector processing on NEC SX-9, the sustained performance degrades. The reason is that SX-9 vector extensions that include mask operations are not valid for no automatic optimization environments. As a result, a branch condition remains in a loop body and it causes the performance degradation.

Figure 7 shows the normalized execution times of each optimization and their combinations with the most advanced automatic optimizations by the compilers. This figure shows that the invariant-code motion in NEC SX-9 also reduces the sustained performance. By moving the calculations to outside of the loop, these calculations are performed by the scalar units in the vector processor. As the scalar unit in NEC SX-9 is not fast as the vector units, the execution time becomes long. In spite of the reduction in the redundant calculations by the invariant-code motion, the execution time becomes long when a scalar processing is performed instead of a vector/SIMD processing.

From Fig. 7, it is clearly shown that uses of the mask operations drastically degrade the performance in the scalar systems while it achieves higher performance in the vector system In order to let a compiler to use mask operations and to perform efficient SIMD processing, the conditional branch using a mask array that in advance stores whether branches are true or not remains in a loop body. However, the compilers in these scalar processors cannot generate an expected code that uses SIMD operations with mask operations. Furthermore, automatic optimizations by

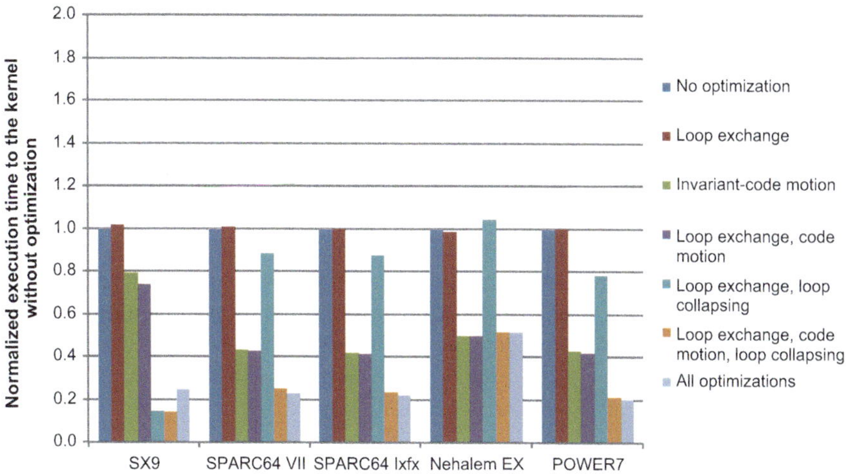

Fig. 6 Speedup by the combinations of the optimization methods without any automatic optimizations

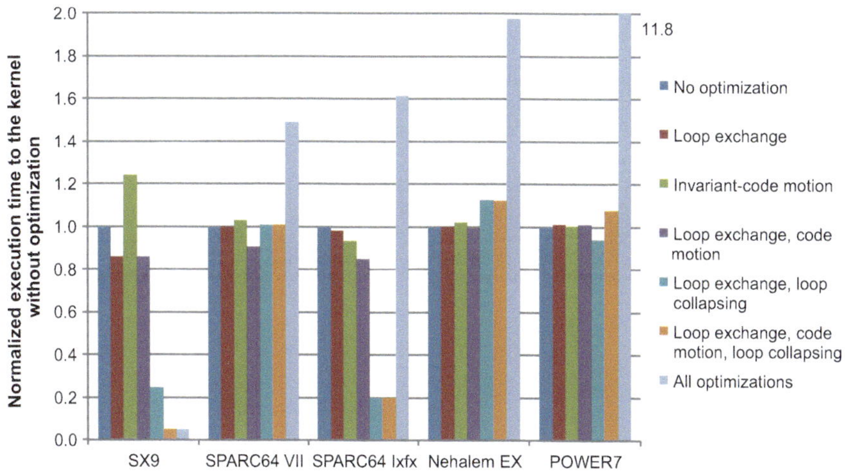

Fig. 7 Speedup by the combinations of the optimization methods with the most advanced automatic optimizations

the compilers cannot be applied to the code. By applying uses of mask operations, the automatic loop unrolling cannot be applied. This causes the severe performance degradation. Therefore, the optimization method such as uses of mask operations that prevents automatic optimizations by compilers should carefully be treated.

5 Conclusions

In order to analyse the performance differences of the optimization methods among multiple HPC systems, this paper evaluates the performances of several optimization methods that are widely utilized for HPC codes. From the performance evaluations that compare the sustained performances between the original kernel and the optimized kernel on each HPC system, the optimization method sometimes degrades the sustained performance. Although an invariant-code motion basically improves the sustained performance, the sustained performance may be reduced by a scalar processing instead of a vector processing in the vector supercomputing system. The slower scalar units than vector units in a vector processor sometimes cause performance degradation. Also, uses of mask operations that enhance vector and SIMD processing of conditional branches using precomputed results might degrade the sustained performance. Since the codes become complicated by applying the optimization, a compiler cannot apply automatic optimizations and the sustained performance is degraded. As the automatic optimizations by a compiler can improve the sustained performance, the optimization method that prevents automatic optimizations should be avoided. In conclusions, the optimization method should be applied by carefully considering the performances of the optimization and a HPC system.

Acknowledgements Authors would like to thank Information Initiative Center, Hokkaido University, Cyberscience Center, Tohoku University, Information Technology Center, University of Tokyo, and Information Technology Center, Nagoya University for the supercomputing resources used for the performance evaluation.

This research was partially supported by Grant-in-Aid for Scientific Research (S) #21226018, Grant-in-Aid for Scientific Research (B) #25280041, and Core Research of Evolutional Science and Technology of Japan Science and Technology Agency (JST CREST) "An Evolutionary Approach to Construction of a Software Development Environment for Massively-Parallel Heterogeneous Systems".

References

1. Top 500 supercomputers sites. http://www.top500.org/.
2. Thomas H. Dunigan Jr., Jeffrey S. Vetter, James B. White III, and Patrick H. Worley. Performance evaluation of the cray x1 distributed shared-memory architecture. *IEEE Micro*, 25(1):30–40, January 2005, http://dx.doi.org/10.1109/MM.2005.20.
3. Yukihiro Hasegawa, Jun-Ichi Iwata, Miwako Tsuji, Daisuke Takahashi, Atsushi Oshiyama, Kazuo Minami, Taisuke Boku, Fumiyoshi Shoji, Atsuya Uno, Motoyoshi Kurokawa, Hikaru Inoue, Ikuo Miyoshi, and Mitsuo Yokokawa. First-principles calculations of electron states of a silicon nanowire with 100,000 atoms on the k computer. In *Proceedings of 2011 International Conference for High Performance Computing, Networking, Storage and Analysis*, SC '11, pages 1:1–1:11, 2011, http://doi.acm.org/10.1145/2063384.2063386.
4. Abtin Rahimian, Ilya Lashuk, Shravan Veerapaneni, Aparna Chandramowlishwaran, Dhairya Malhotra, Logan Moon, Rahul Sampath, Aashay Shringarpure, Jeffrey Vetter, Richard Vuduc, Denis Zorin, and George Biros. Petascale direct numerical simulation of blood flow on 200k

cores and heterogeneous architectures. In *Proceedings of the 2010 ACM/IEEE International Conference for High Performance Computing, Networking, Storage and Analysis*, SC '10, pages 1–11, 2010, http://dx.doi.org/10.1109/SC.2010.42.
5. Takashi Shimokawabe, Takayuki Aoki, Tomohiro Takaki, Toshio Endo, Akinori Yamanaka, Naoya Maruyama, Akira Nukada, and Satoshi Matsuoka. Peta-scale phase-field simulation for dendritic solidification on the tsubame 2.0 supercomputer. In *Proceedings of 2011 International Conference for High Performance Computing, Networking, Storage and Analysis*, SC '11, pages 3:1–3:11, 2011, http://doi.acm.org/10.1145/2063384.2063388.
6. Takashi Soga, Akihiro Musa, Youichi Shimomura, Ryusuke Egawa, Ken'ichi Itakura, Hiroyuki Takizawa, Koki Okabe, and Hiroaki Kobayashi. Performance evaluation of nec sx-9 using real science and engineering applications. In *Proceedings of the Conference on High Performance Computing Networking, Storage and Analysis*, SC '09, pages 28:1–28:12, 2009. http://doi.acm.org/10.1145/1654059.1654088.

Power Consumption of Kernel Operations

Dmitry Khabi and Uwe Küster

Abstract A modern Petascale System consists of millions of different components, which consume a huge amount of energy. The power rating of each component depends on the type of the current instructions, executed on cores, memory controllers, network units and other various components. There are a lot of influences and complicated dependencies between the software, environment and the energy consumption. The objective of this work is to identify and understand the energy consumption of processors and memory in the consideration of kernel operations. Another important goal is to develop the methodology by which the developers and users could estimate the energy consumption of the different algorithms on different systems with minimal effort and satisfying accuracy.

1 Introduction

1.1 Power Consumption of Computational Node

To review the assertion that the processor and the memory consume a substantial part of the energy, we implemented a test bed that we use to measure the electrical power not only of whole computational node but also of its hardware components. The workstation with Intel®Xeon®Processor E5-2687W was selected for the analysis. The technical description of the workstation is in Table 1. The hardware components are divided into the six groups `Motherboard`, `CPU+RAM`,

D. Khabi (✉) · U. Küster
High Performance Computing Center Stuttgart (HLRS), Nobelstraße 19, D-70569 Stuttgart, Germany
e-mail: khabi@hlrs.de; kuester@hlrs.de

M.M. Resch et al. (eds.), *Sustained Simulation Performance 2013*,
DOI 10.1007/978-3-319-01439-5_3,

Table 1 Technical description of the workstation

Components	Description
Motherboard	Supermicro®X9SRA Single Socket R (LGA 2011)
Processor	Sandy Bridge E5-2687W (8 cores, 20 M L3-Cache, 3.10 GHz)
Memory	4 × Kingston®Server Premier 4 GB Module–DDR3 1,600 MHz ECC
Video	Simple graphic card (HD5450, 1 GB DDR-3, passive cooling)
Power supply	Antec®EarthWatts EA-450 Platinum 450 W
Hard disc	Toshiba®DT 01ACA100 (1 TB, 72,000 rpm, 32 M Buffer)
CPU-Fan	Be Quiet!®DARK ROCK 2, 135 mm
Chassis Fans	(×2) 200 mm Fans (×1) 120 mm Fans
OS	Scientific Linux release 6.3 (Carbon), kernel version 2.6.32

`Power supply`, `Hard disk` and `Fans`. The different computational loads were exerted by three types of tasks:

- `no load`: No load on the workstation except produced by OS Jitters
- `Add`: Sum of the elements of two arrays (see Sect. 1.3 for more details)
- `copy files`: Copy of a large file from one directory to the one other

We changed not only the type of the task and its size to activate the different hierarchy levels of memory, but also the frequency of the processor. The processor was set either to the lowest or the highest possible frequencies. The power rating of the various components of the workstation for different problems is shown in Fig. 1. The energy consumption of the processor and memory depends on the number of active cores as well as the frequency. Another factor is the active hierarchy level of memory. As you can see the processor together with memory modules are the most significant power consumer.

1.2 Power Measurement

In order to determine the energy consumption of an electrical device, one must know its electrical power and time of its work. The time of the calculation can be easily obtained from the performance of the kernel operations. The electric power (`Watt`) is given by the product of applied voltage (`V`) and the electric current (`A`) in a `DC` circuit. The `V` and `A` are measured at the power connectors on the motherboard. The Fig. 2 shows the schematic diagram of the applied power measurements. The high precision shunts resistors $R = 0.01\ \Omega$ are placed in series with the different hardware components so that the current flows through these. The analog-digital (`A/D`) converter records the potential drop at the shunts in time at high frequency (up to 100 kHz). The `A/D` converter is installed in the additional PC (Host). The recorded values can be easily converted into the electrical power and stored on the hard disc. The voltage level (`V`) should be also measured for the

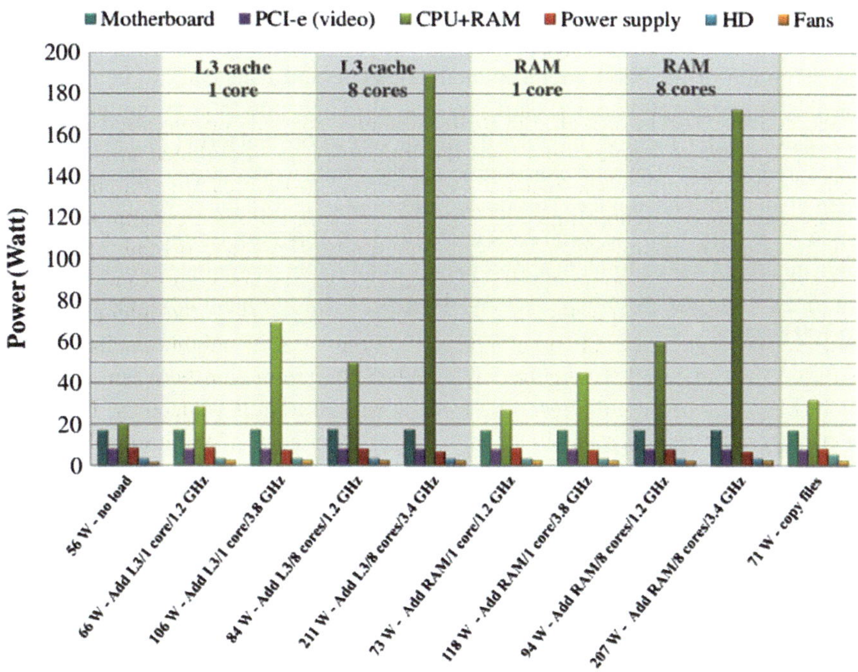

Fig. 1 The electric power of the workstation components (see Table 1) under various loads

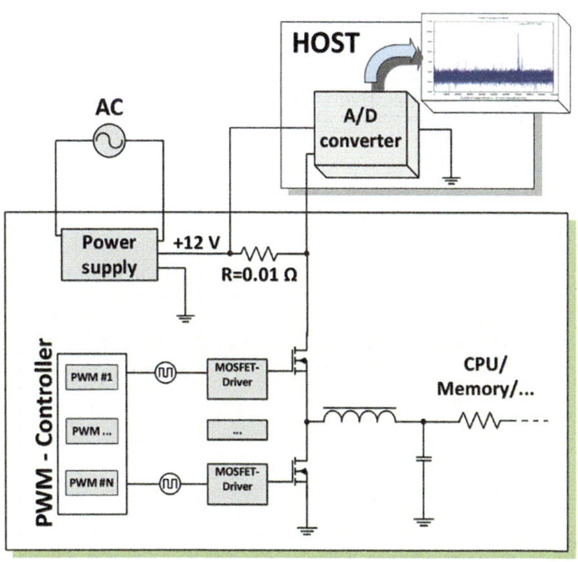

Fig. 2 Schematic diagram for power measurement of processor and memory modules

higher accuracy, because a voltage drop of some percent may occur even when the current of few amperes is passed through the wire. Unfortunately, the power consumption of the CPU, memory modules, PCIe and Fans can not be separated on the Supermicro®motherboard. We calculate the power consumption of the processor and the memory modules by subtracting a constant term (10.5 W) from the measured values.[1] The total consumption of the whole workstation is measured with help of an additional device. This device allows us to record the power consumption in an `AC` circuit with the same `A/D` converter (see appendix for more details). The high quality of our equipment allows to perform reliable power measurements under various conditions[2] with high accuracy ($\epsilon_{relative} < \pm 1.5\,\%$) and time resolution (up to 100 khz).

1.3 Kernel Operations

In this paper we address four kernel operations that have been considered among many other in our work:

- `Add`: Sums of the elements of two arrays that stored in the third array:

$$a_i = b_i + c_i; 0 \leq i < length \tag{1}$$

- `Dot product`: Computational of scalar product:

$$dot+ = b_i \times c_i; 0 \leq i < length \tag{2}$$

- `Load AVX`: Load the values from cache or memory to the CPU AVX registers by using the stream instruction `mm256_load_pd`:

$$a_i -> AVX; \leq i \leq length \tag{3}$$

- `Store AVX`: Store the values from CPU AVX registers to cache or memory by using the stream instruction `mm256_store_pd`:

$$AVX -> a_i; \leq i \leq length \tag{4}$$

[1]We do not run any calculations on the weak graphics card and the fans run at a constant speed. We assume that the power consumption of these components is constant. This was also confirmed with a series of extra tests.

[2]We vary the number of active cores, frequency and hierarchy levels of memory, on that the processor operates.

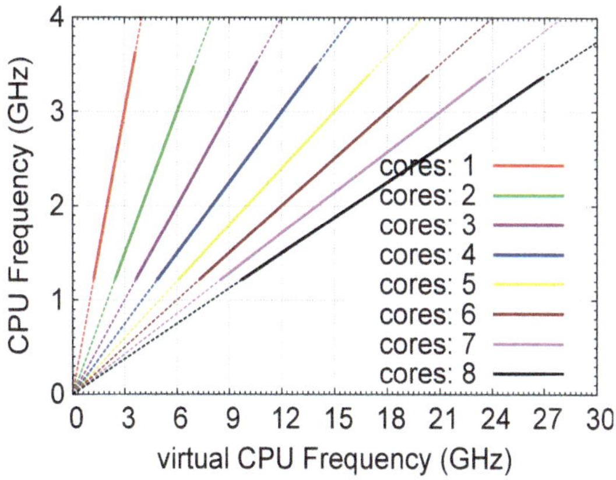

Fig. 3 Dependencies of Frequency and Virtual Frequency of CPU. The maximal virtual frequency of 27.3 GHz is achieved with eight cores in Turbo mode

Depending on the size of the arrays the different hierarchy levels of memory (L1, L2, L3 or RAM) can be activated. The size of the arrays is controlled via variable `length`. It should be noted, though the processed data fit in cache completely some memory related instructions will be nevertheless executed by the processor! The cache coherency policy is in most cases `write-back` i.e. by `write-back` the data is first updated in the cache, and after that the memory controller triggers a memory update. Furthermore, since the cache policy of the Intel processors is `inclusive`, some activity will be registered by other hierarchy levels of memory even if the data is changed in L1 cache, that is marked as exclusive due to the coherency protocol. The intrinsic functions and loop unrolling were used in the implementation of the kernel operations; This gives us a better opportunity to control the hardware. The kernel operations are parallelized with OpenMP. For setting of different frequencies we use the package “lib cpufreq” that is available in most distribution of OS Linux.

1.4 Virtual CPU Frequency

The Intel processor Sandy Bridge gives the user the possibility to operate its cores with different frequencies. The cores can also operate at different frequencies simultaneously. We don’t use this feature; All cores operate with the same frequency in all examples, that refers in this paper. For some diagrams we use the metric `virtual CPU frequency` instead of `CPU frequency`. The virtual frequency depends on the number of active cores and their frequency. For example if two cores are used for the calculation and each runs with 1.2 GHz, the virtual frequency is equal to 2.4 GHz. The Fig. 3 shows the dependencies between them. If all cores operates in Turbo mode, than the frequency of each core is 3.4 GHz. If only one core is active Turbo frequency is equal to 3.8 GHz.

2 Electric Power of Processor (Watt, GHz)

As mentioned above, we need to know the performance of the kernel operations and the processor power in order to calculate the energy consumption. In this section we present electric power approximation model. The benefit of these in comparison to the existing models is that the values can be obtained with high accuracy via the analytic formula. And moreover the constants in the formula reflects different properties of the processors.

2.1 Power Approximation

According to the Intel White Paper for the Pentium®M processor, that supports Enhanced Intel SpeedStep Technology, the electric power of it is approximately proportional to its frequency, and to the square of its voltage [1]:

$$Power = CV^2 f \tag{5}$$

Where `C` is the capacitance, `V` is the actual voltage and `f` is the actual frequency. The direct use of this formula is not possible in practice. A modern processor changes not only its voltage and frequency, but also the capacitance depending on the executable instructions and hardware configuration. In addition, the various components of the processor operate at different frequencies. For example the L3 cache and its ring operate always with the nominal frequency (3.1 GHz for `E5-2687W`). Instead of the formula (5), we decided to check on the suitability of a new formula for the approximation of its electric power depending on the frequency.

$$Power(f)_{lp} = \alpha_{0\,lp} + \alpha_{3\,lp} \times f^{\rho} \tag{6}$$

Where the constants α_{0lp} and α_{3lp} are to calculate for the different hierarchy levels of memory (index l), number of active cores (index p) and selected kernel operation. The constants α_{0lp} and α_{3lp} is greater than zero and the exponent ρ is greater as one. The constant α_0 should be equal to the consumption of the processor's components, which are independent of the variable frequency. Note, the calculations at the different hierarchy levels of memory may well affect the both constants. In our case the power of the memory modules should be also included in α_0. The term $\alpha_3 \times f^{\rho}$ reflects the changes in the power consumption depending on the actual frequency. The exponent ρ and the constant α_3 indicates the speed of the increasing of the power and depends on the processor type, quality of the chip, required voltage for different frequencies and some other factors. The next question is, how do we calculate the constants and the exponent of (6)? Using our test bed we can measure the electric power of the processor and memory modules during the execution of the kernel operations. We recorded the electric power of the processor and memory modules when the kernel operations rans on up to eight cores with all possible frequencies and with data in different hierarchy levels of memory (L1, L2, L3, RAM). After this we composed `N` linear systems ($N = 8\,cores \times 4\,memory\,levels$) each with `n` equations and two unknowns. Where `n` = 16 is the number of possible frequencies of the processors (the Intel processor E5-2687W can operate by 16 different frequencies: $f_0 \ldots f_{15}$):

$$\begin{vmatrix} 1 & f_0^{\rho} \\ 1 & f_1^{\rho} \\ \cdots & \cdots \\ 1 & f_{15}^{\rho} \end{vmatrix} \times \begin{vmatrix} \alpha_{0\,lp} \\ \alpha_{3\,lp} \end{vmatrix} \approx \begin{vmatrix} power_{0\,lp} \\ power_{1\,lp} \\ \cdots \\ power_{15\,lp} \end{vmatrix} \tag{7}$$

The values $power_0 \ldots power_{15}$ are the measured electric power. These linear systems are over determined. By using the `QR` algorithm for a least squares approach, the constants α_0 and α_3 can be found for a selected ρ for each memory level l and number of active cores p so that the mean error of the approximation is equal to zero and the error is minimal. For the choice of the exponent ρ let us define the absolute error ϵ_{abs} of the approximation:

$$\epsilon_{min\,abs} = \min_{i\,l\,p}(power_{i\,lp} - (\alpha_{0\,lp} + \alpha_{3\,lp} \times f_i{}^{\rho})) \tag{8}$$

$$\epsilon_{max\,abs} = \max_{i\,l\,p}(power_{i\,lp} - (\alpha_{0\,lp} + \alpha_{3\,lp} \times f_i{}^{\rho})) \tag{9}$$

$$i \in (1.2\,GHz, 1.3\,GHz, \ldots);\ l \in (L1, L2, L3, RAM);\ p \in (1\,core, ..., 8\,cores) \tag{10}$$

$$\epsilon_{max\,abs} \geq 0 \tag{11}$$

$$\epsilon_{min\,abs} \leq 0 \tag{12}$$

$$\epsilon_{abs} = \epsilon_{max\,abs} - \epsilon_{min\,abs} \tag{13}$$

The absolute error ϵ_{abs} is the sum of the two maximum differences (negative and positive) between measured and approximated power over all cores and four levels of the memory hierarchy.

2.2 *Power of Kernel Operations*

The best approximation (within the meaning of ϵ_{abs}) for the kernel `Add`, can be achieved with the exponent $\rho = 2.42$. In this case, the absolute error of the approximation is 8.021 `Watt`.[3] The Fig. 4 shows the electric power of the Intel

[3]The best approximation for the same kernel operation Add on Sandy Bridge Intel i5-2500 (6 M Cache, up to 3.7 GHz, 4 cores) is by $\rho = 2.7$.

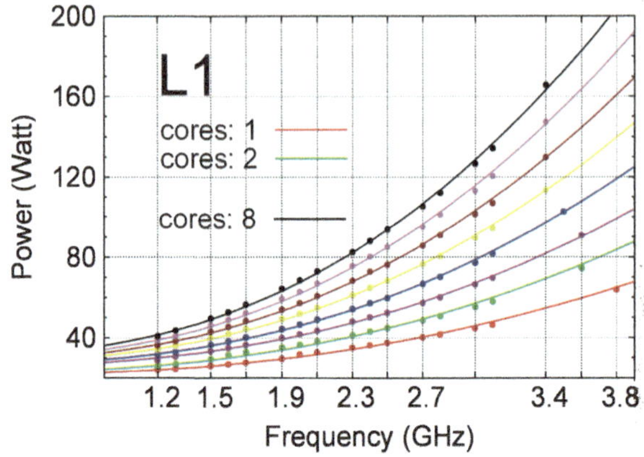

Fig. 4 Electric power of kernel operation Add on the Intel processor E5-2687W, when the data is in L1 cache

Table 2 The power approximation parameter α_0 and α_3 for several cases

Cores	Mem	Add	α_0	α_3	ρ	$\epsilon_{\tilde{abs}}$	Load	α_0	α_3	ρ	$\epsilon_{\tilde{abs}}$
1	L1	–	21.3	1.7	2.42	5.23	–	21.3	1.4	2.51	6.92
8	L1	–	30.8	6.8	2.42	5.23	–	29.9	5.1	2.51	6.92
1	L2	–	21.3	1.7	2.42	5.30	–	21.4	1.5	2.51	7.25
8	L2	–	30.8	7.0	2.42	5.30	–	30.5	5.6	2.51	7.25
1	L3	–	25.6	1.8	2.42	6.47	–	24.2	1.46	2.51	7.13
8	L3	–	38.8	7.8	2.42	6.47	–	38.0	6.2	2.51	7.13
1	RAM	–	25.5	0.8	2.42	7.89	–	25.2	0.62	2.51	9.69
8	RAM	–	50.8	3.4	2.42	7.89	–	44.1	6.0	2.51	9.69

processor E5-2687W with four memory modules (4 × 4 GB) depending on the frequency, when the kernel Add was run on up to 8 cores and the data was in level 1 cache.

The measured values are marked with the dots. The curves shows the approximated power according to the formula (5) with $\rho = 2.42$ for one (red) up to eight (black) cores. The other curves colors can be easily assigned to the number of active cores. The power consumption increases with the increase of the frequency and number of active cores. For some l and p and the kernel operations Add and Load the constants α_0 and α_3 are listed in the Table 2. Additionally the values for $\epsilon_{\tilde{abs}}$ are listed. The $\epsilon_{\tilde{abs}}$ indicates the absolute error only over the corresponding hierarchy level of memory (for all i and p; l is fixed) in contrast to the ϵ_{abs} that indicates the error over all memory levels: $\epsilon_{abs} \geq \epsilon_{\tilde{abs}}$ (for all i , l and p). As expected, the constant α_0 is about the same for both kernels. The difference is only considerably if the data are in RAM. Because the throughput of instructions of kernel Load is

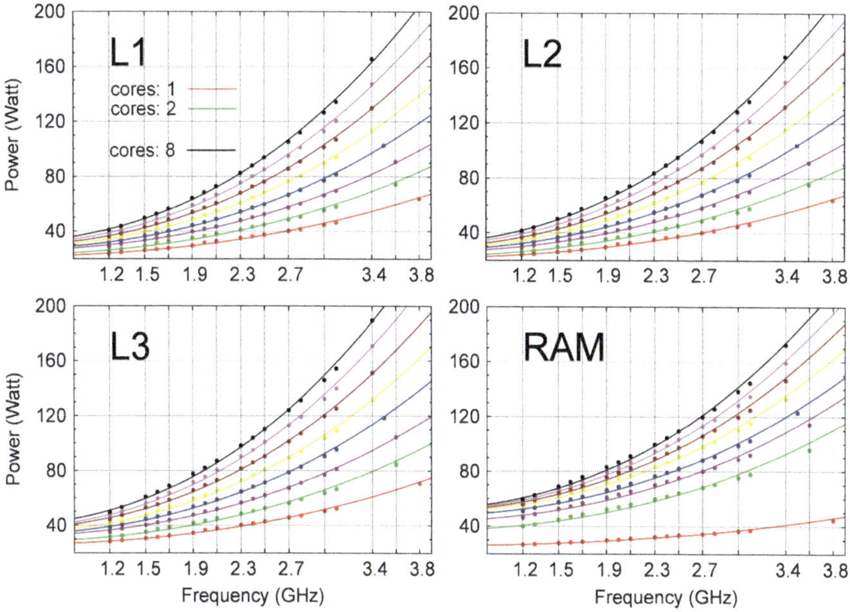

Fig. 5 Electric power of kernel operation Add on the Intel processor E5-2687W, when the data is in L1, L2, L3 caches and in RAM

higher, the power consumption increases faster. This fact is indicated by ρ. The Fig. 5 shows the measured and approximated power for all hierarchy levels of memory.

You can see that the α_0 and α_3 differ only slightly for L1 and L2. The constant term $\alpha_{3\,RAM}$ increases greatly when at least two cores are active.

3 Performance of Processor (Elements per Second, GHz)

In this section we present performance approximation model which has a high accuracy and is more suitable for our purpose than models as for example described in [3] and [5].

3.1 Performance of Kernel Operations

The Fig. 6 shows the performance of the kernel operation `Add` depending on the level of memory, the frequency and the number of active cores. The metric `Elements per second` [10^9] indicates how many billions of array elements

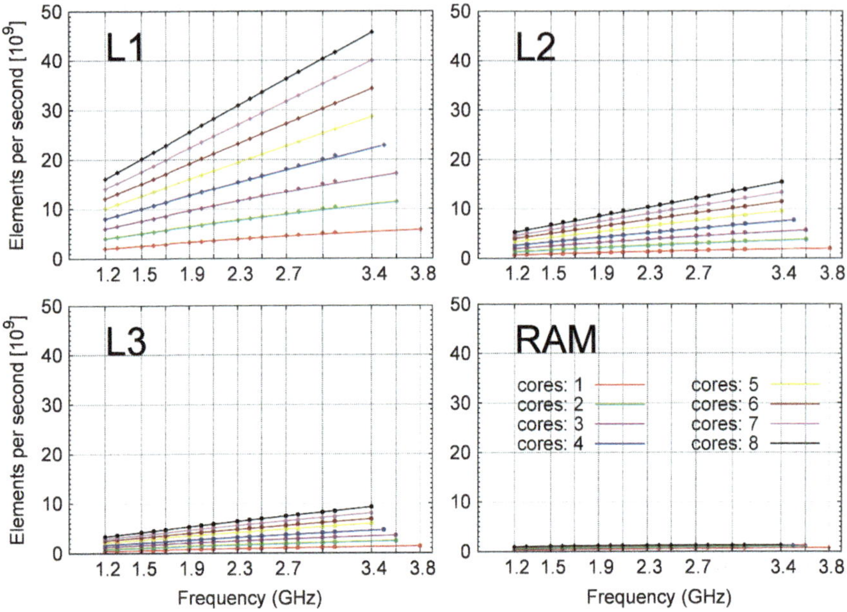

Fig. 6 Performance of operation Add on the Intel processor E5-2687W, when the data are in L1, L2, L3 caches and in RAM

a_i were calculated in a second. 10^9 elements, not one, is used in the unit as the base for the number of elements. We use the same colors as in the power diagrams to distinguish the number of active cores (red for one core , black for eight cores). The measured performance is marked with the dots. The curves shows the approximation of the performance, which is described in the next Sect. 3.2. In contrast to the memory bandwidth, the cache bandwidth is directly related to the frequency of the CPU and increases linear, because L1 and L2 caches run at the same frequency as theirs core. The bandwidth of L3 is almost three times smaller than the nominal one. This is explained by the fact that the newest Intel processors provide the rings in L3 cache: Each core is connected to the local segment of the L3 cache. The adjacent segments are connected via ring bus. If the data for the calculation is stored in the owned segment of L3, it can be immediately delivered to the core. Otherwise the required data will hop through the ring until it reaches the core. The data in L3 cache is cyclically distributed across all segments, which are connected with each other over the ring bus.

It is well recognizable that the memory performance is significantly lower than the cache performance. According to the article "Intel's Haswell CPU Microarchitecture" [2] and slides presented at Intel Delveloper Forum 2012 [4], the nominal bandwidth on Sandy Bridge (E5-2687W) of L1 cache is 1216 GiB/s and of L2 and L3 is equal to 811 GiB/s. The bandwidth of DDR3 RAM is only 47.7 GiB/s. The electrical power, however not substantially differs whether the data processing

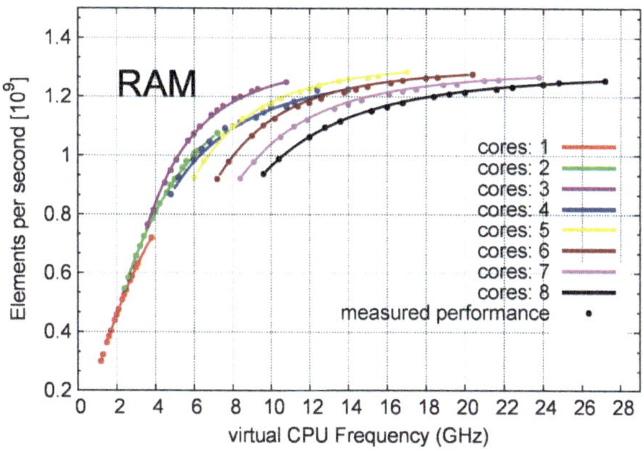

Fig. 7 Performance approximation of operation Add on the Intel processor E5-2687W according to the formula (15), if the data are in RAM

takes place in memory or in cache (see Fig. 5). Before we begin with the direct comparison of energy consumption, we have to approximate the performance data with an analytic formula.

3.2 *Performance Approximation*

The performance of the cache increases generally linearly with the frequency. Only in turbo mode by few active cores you can see a small deviation from the linear ratio. The reason is not very clear for us. One reason may be the fact that L3 cache and memory controller running at constant frequency. This could explain smaller delay when the data in cache and memory are synchronized. Much interest is the ratio of the performance in RAM. The Fig. 7 shows the performance of the kernel operation `Add` if the data are in RAM. The maximal bandwidth is achieved with five cores. However, the difference to the performance achieved with three and six core is within the measurement accuracy. The average value of ten independent measurements differs by about 1 % of the maximum and minimum. The exception is the performance data for the four cores. In this case, the deflection reached up to 3 %, as can be seen in the diagram; This is the reason, why the blue dots are more scattered than the others.

The linear approximation of the performance as function of frequency shows good approximation in case of cache:

$$perf(f)_{lp} = \gamma_{0\,lp} + \gamma_{1\,lp} \times f \tag{14}$$

But if the data are stored in RAM, the error is too large. The unsymmetrical utilization of memory channels relative to the number of cores contributes

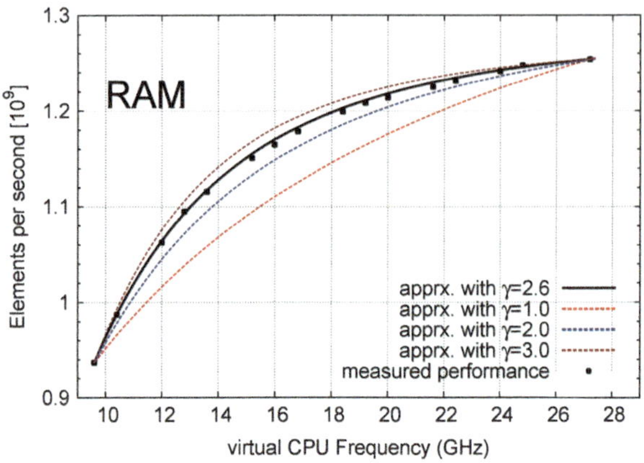

Fig. 8 Performance approximation of operation Add for eight cores according to the formula (15). The curves show different results of the approximation in depending on the exponent λ.

additional complexity to the approximation of the performance. Known models of the performance don't allow to improve their quality of the approximation to a sufficient extent. The roofline model [3] provides realistic expectations of performance, that can be achieved with regard to the memory bandwidth limitation and arithmetic instructions throughput. But the above mentioned aspects are not considered sufficiently. The same can be said about the ECM performance model described in [5]. We are still in the process of finding a suitable approximation formula, which is quit simple and not only approximates the performance with minimum error, but also reflects the different processes ongoing in the CPU. The formula should be suitable not only for the kernel operations but also for more complex algorithms. But for present purposes we use the formula, that approximates the performance of the kernel operations quit well. The selected formula of the performance and its important properties are as follows:

$$perf(f)_{lp} = \frac{\gamma_{max\,lp}}{\gamma_{lp} + \frac{1-\gamma_{lp}}{f^{\lambda_{l\,p}}}} \tag{15}$$

$$\lim_{f\to 0} perf(f) = 0 \tag{16}$$

$$\lim_{f\to\infty} perf(f) = \frac{\gamma_{max}}{\gamma};\ \gamma \neq 0 \tag{17}$$

The (16) expresses that by reducing of the frequency the performance goes to zero. The expression (17) means the increasing of the frequency enhances the performance, that is limited by the maximum bandwidth. The Fig. 8 shows the approximation of the performance for different exponents λ. The error of the performance approximation for eight cores $\epsilon_{\tilde{abs}}$ is minimal by $\lambda = 2.6$ and equal to 7^{-3} (less than 1 % of the achieved performance).

Table 3 The performance approximation parameters γ_{max}, γ and λ for several test cases

Cores	Mem	Add	γ_{max}	γ	λ	$\tilde{\epsilon_{abs}}$	Load	γ_{max}	γ	λ	$\tilde{\epsilon_{abs}}$
1	L1	–	1.64	0.14	1.33	0.28	–	3.87	0.14	1.34	0.73
8	L1	–	13.41	0.00	1.00	0.04	–	31.70	0.00	1.00	0.08
1	L2	–	0.54	0.14	1.32	0.08	–	1.53	0.11	1.20	0.32
8	L2	–	4.37	0.00	1.01	0.23	–	12.16	0.00	1.03	0.67
1	L3	–	0.37	0.06	1.12	0.01	–	1.08	0.04	1.07	0.04
8	L3	–	2.76	0.03	1.05	0.06	–	8.54	0.04	1.08	0.25
1	RAM	–	0.26	0.14	1.03	0.00	–	0.83	0.19	1.16	0.04
8	RAM	–	0.80	0.63	2.60	0.01	–	2.36	0.36	2.23	0.32

Some performance approximation parameters for kernel operations `Add` and `Load` are listed in Table 3. The performance approximation (15) has high accuracy. Just in case if the data are in the cache and less cores are active, linear approximation (14) has better accuracy for some frequencies. How it affects the results, can be seen in the next section.

4 Energy Consumption of Kernel Operations (Joule per Element)

Once we have calculated the power and the performance of the kernel operations, we can calculate the dependence between energy consumption of the processor and memory modules and the achieved performance. Note, the performance increases due to the increase of frequency.

4.1 *Dependencies of Energy and Performance*

Figure 9 shows the energy consumption in nanojoules for the calculation of one element a_i of the kernel operation `Add`. The dots mark the measured values. The curves shows the approximation of the energy. By the dashed gray curves we used the linear approximation of the performance. By the bold curves we used the approximation with function (15). The bold curves are defined on the performance area that can really be achieved with the processor (over the setting of different frequencies). The bold curves are continued with the dashed colored curves. These extrapolate the consumption at higher frequencies that cannot be achieved by the processor. The extrapolation error is of course unknown.

As you can see, the approximated values match with those measured very well in the case of calculations on four and more cores. By calculations with less than four cores the energy consumption's characteristics remain, although there is relatively

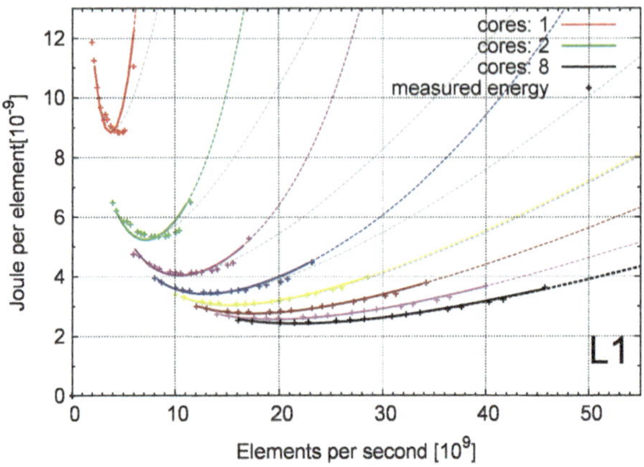

Fig. 9 The curves shows how many nanojoules are consumed by the processor and memory modules for the calculation of one element a_i with the corresponding performance. The data are in L1 cache

small deviations. The greatest amount of energy consumed by the calculations on a single core. This is not surprising, since the processor and the memory has many components that consume a constant amount of energy regardless of the number of active cores (compare α_0). The smallest amount of energy consumed by the calculation with eight cores, while increasing of the performance is not so strongly affect the energy consumption.

Figure 10 shows the energy consumption in nanojoules depending on the performance for all hierarchy levels of memory.[4] The axes are scaled differently. What is remarkable is a big difference between different cases:

- If the memory is not active there are big advantages of increasing of frequency and activation of more cores.
- If the memory is active, there are no apparent advantages of increasing of frequency to the highest possible. The performance increases very slow and the energy consumption rises very quickly. The next significant difference is that the calculation on all eight cores is not optimal, neither for the performance nor for the consumption of energy.

The Fig. 11 shows the consumption of operation `Add` with one and eight active cores for all hierarchy levels of memory. The X and Y axis are in base 2 logarithmic scale. The differ between the left and right diagrams is, that the consumption of the rest of the system was considered by adding of 30 W to the constant term α_0 (see formula (6)). In doing so not only the energy amount was increased for

[4] If we compare the obtained values to the data from the article [6] we see that the energy consumption of Sandy Bridge (E5-2687W) and of the previous generation Westmere (Intel XEON X5670) are within the same order of magnitude.

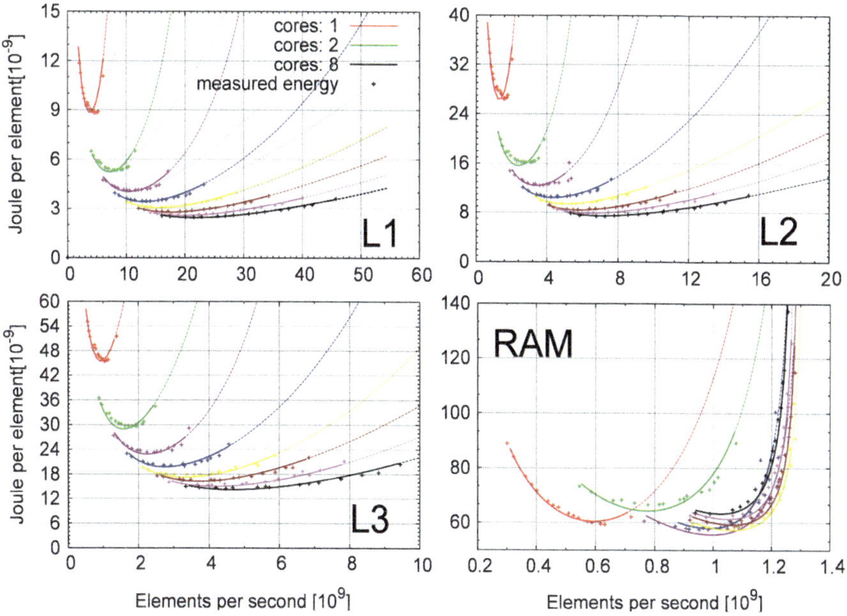

Fig. 10 The curves shows how many nanojoules are consumed by the processor and memory modules for the calculation of one element a[i] with the corresponding performance

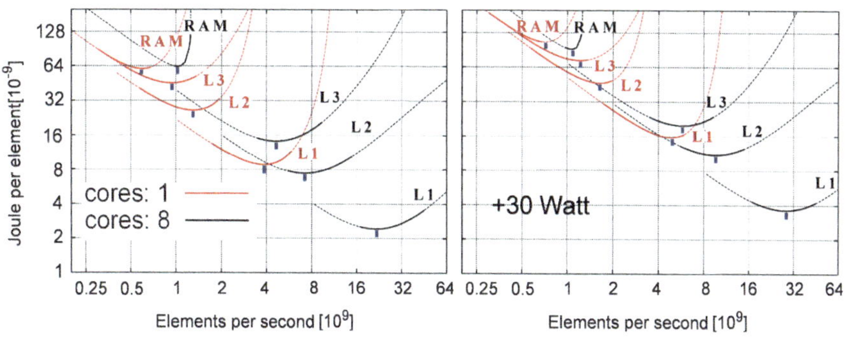

Fig. 11 The *left* diagram shows the energy consumption in nanojoules of the processor and memory modules for computing of one element of the operation Add. The *right* diagram shows the energy consumption of the processor, memory modules and the rest of the system

the calculation of one element, but also the steepness of the curves was changed; The minimum values (marked with vertical short lines) of the approximation were shifted slightly to the right. This means that the optimum frequency (within the meaning of the energy) became higher. Note, we assume here that the rest of the

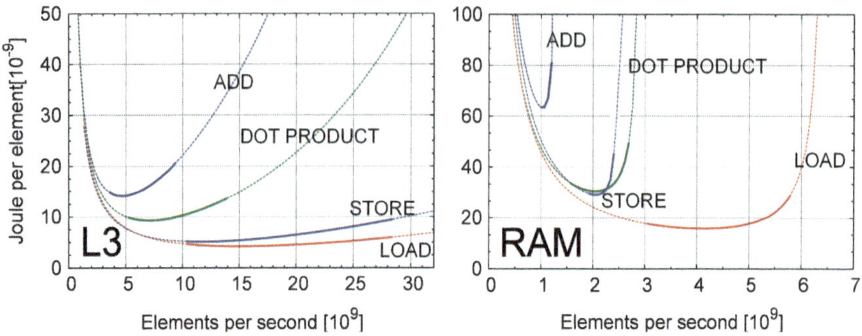

Fig. 12 Energy consumption in nanojoules of kernel operations Add, Dot product, Load, Store on the Intel processor E5-2687W, when the data are in L3 caches and in RAM and all eight cores are active

system consumes 30 W. In reality, the additional consumption is not necessarily constant, because at higher consumption, the power supply has higher efficiency. Also, the cooling system may consume more energy.

4.2 Compare Kernel Operations

Now we can compare the energy consumption of four different kernels, that are described in Sect. 1.3. Figure 12 shows how many nanojoules are needed to carry out these operations. The most expensive and slow is the operation `Add`. The other operations consume less energy and are faster. In case of the calculation in level 3 cache the operations `Store` and `Load` don't differ greatly. But in case of the calculation in memory the operation `Store` needs much more energy and is slow; Before the data can be stored it must be read from the memory. The consequence is that the store access requires two expensive streams from and to the memory.

In case of the calculation in memory the different kernel operations have different benefits from the increasing of the frequency. While the optimum frequency of `Add` is very low, the optimum frequency of the others operation is higher. It shows, that different algorithms has its own optimum frequency.

5 Outlook

To approximate the power of processor and memory modules we have used the formula

$$Power(f)_{lp} = \alpha_{0\,lp} + \alpha_{3\,lp} \times f^{\rho} \tag{18}$$

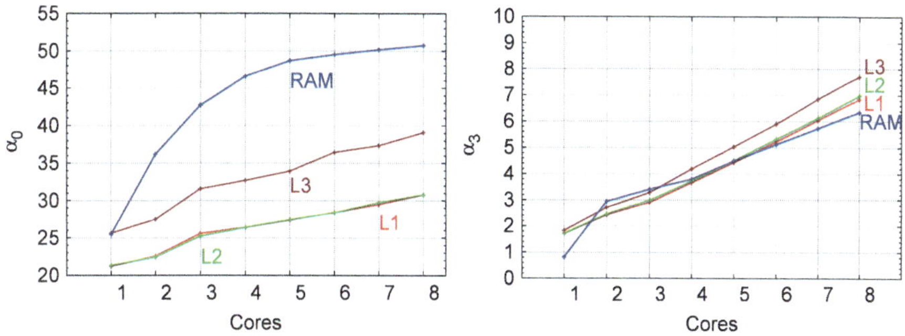

Fig. 13 The constants α_0 and α_3 depending on the number of active cores and the hierarchy level of memory

The constants α_0 and α_3 must be computed not only for different algorithms but also for different hierarchy levels of memory l and number of active cores p. There are total $2 \times |l| \times |p|$ coefficients for all possible cases. If we consider the dependence of the coefficients for example on p, we can greatly reduce the number of coefficients. Figure 13 shows this dependence for the kernel operation `Add`. As you can see the constants α_0 and α_3 are mostly similar in case of `L1` and `L2`. If the calculation is in cache the constants depends practically linear on the number of cores. The other interesting fact is that the difference between `L3` and `L1` and `L2` for α_0 remains constant. Unfortunately, the behaviors of α_0 and α_3 for the memory are more complicated. We are currently considering various approaches to solving this problem. Furthermore, we expand the range of analyzing kernel operations to more complex algorithms.

If at the present time the dynamic switching of different frequencies has a number of barriers, in the future it is possible that this will be one of the keys to more efficient use of energy than is common today.

Appendix

The scheme of the device, that allows us to record the power consumption in an `AC` circuit with the `A/D` converter is shown in Fig. 14

The recorded voltage and current profiles of a workstation as a function of time are shown in Fig. 15.

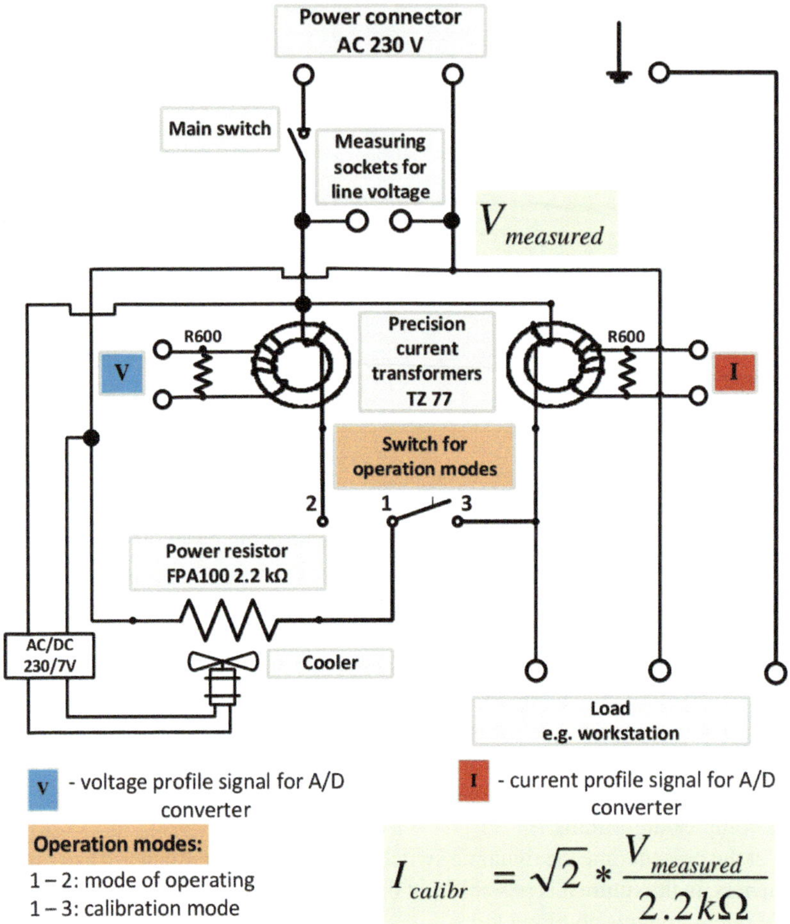

Fig. 14 Schematic diagram of the auxiliary device for the measurement in AC circuit

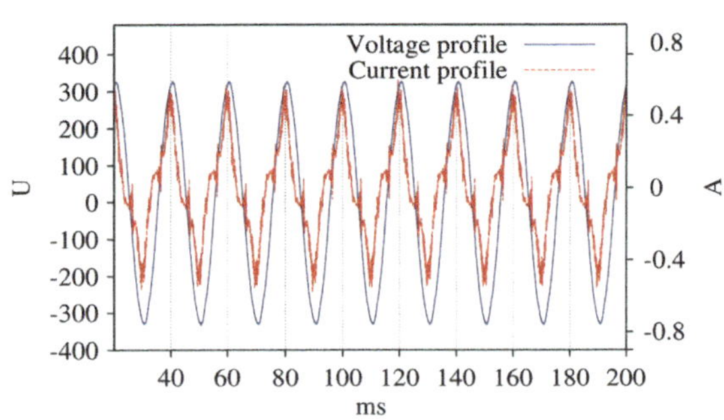

Fig. 15 The voltage and current characteristics. The measurement was performed with the frequency of 50 kHz. The power in discrete time can be calculated with the formula: $Power = \frac{1}{N}\sum_{n=0}^{N-1} V(n) \times I(n)$, Where N is the number of the measurements

Acknowledgements This work has been supported by the CRESTA project that has received funding from the European Community's Seventh Framework Programme (ICT-2011.9.13) under Grant Agreement no. 287703 and by the ExaSolvers project that has received funding from the German Research Foundation (DFG) as part of the Priority Programme "Software for Exascale Computing–SPPEXA".

References

1. Enhanced Intel SpeedStep Technology for the Intel Pentium M Processor - White Paper. ftp://download.intel.com/design/network/papers/30117401.pdf Intel Corparation (2004)
2. David Kanter, Intel's Haswell CPU Microarchitecture http://www.realworldtech.com/haswell-cpu (November 2012)
3. S. Williams, A. Waterman and D. Patterson. Roofline: An Insightful Visual Performance Model for Multicore Architectures Communications of the ACM, Vol. 52, No. 4. (April 2009), pp. 65–76, doi:10.1145/1498765.1498785
4. Robert Chappell, Bret Toll, Ronal Singhal, Intel Next Generation Micro Architecture Code-name Haswell: New Processor Innovations. Presented at IDF (2012)
5. Markus Wittmann, Georg Hager, Thomas Zeiser, Gerhard Wellein. An analysis of energy-optimized lattice-Boltzmann CFD simulations from the chip to the highly parallel level arXiv:1304.7664 (April 2013)
6. Daniel Molka, Daniel Hackenberg, Robert Schöne and Matthias S. Müller, Characterizing the Energy Consumption of Data Transfers and Arithmetic Operations on x86-64 Processors, Proceeding of the first International Green Computing Conference, (August 2010)

Part II
Frameworks and Libraries for Simulations on New-Generation Computing Systems

Lattice Boltzmann Simulations on Complex Geometries

Simon Zimny, Kannan Masilamani, Kartik Jain, and Sabine Roller

Abstract The need for numerical simulation of fluid flows in highly complex geometries for medical or industrial applications has increased tremendously over the recent years. In this context the lattice Boltzmann method which is known to have a very good parallel performance is well suited. In this publication the lattice Boltzmann solver Musubi which is a part of the end-to-end parallel simulation framework APES is described concerning its HPC performance on two possible applications. The first application is the blood flow through stented aneurysms including a simple clotting model, the second application is the flow of water through an industrial spacer geometry. In both cases, a highly complex geometry with a wide range of spatial scales (μm up to cm) each is used.

S. Zimny (✉)
German Research School for Simulation Sciences and RWTH Aachen University, Schinkelstr. 2a, 52062 Aachen, Germany

University of Siegen, Simulation Techniques and Scientific Computing, Hölderlinstr. 3, 57068 Siegen, Germany
e-mail: s.zimny@grs-sim.de; simon.zimny@uni-siegen.de

K. Masilamani
University of Siegen, Simulation Techniques and Scientific Computing, Hölderlinstr. 3, 57068 Siegen, Germany

Siemens AG, Corporate Technology, CT RTC ENC ENT-DE, Günther-Scharowsky-Str. 1, 91058 Erlangen, Germany
e-mail: kannan.masilamani@uni-siegen.de

K. Jain · S. Roller
University of Siegen, Simulation Techniques and Scientific Computing, Hölderlinstr. 3, 57068 Siegen, Germany
e-mail: kartik.jain@uni-siegen.de; sabine.roller@uni-siegen.de

M.M. Resch et al. (eds.), *Sustained Simulation Performance 2013*,
DOI 10.1007/978-3-319-01439-5_4,

1 Introduction

In recent years, a wide range of applications require highly intensive numerical simulation of fluid flow in complex geometries. An efficient numerical scheme and data structures are required to deploy computation intense large problems on supercomputers and clusters. The lattice Boltzmann method (LBM) which is highly suitable for fluid flow simulations in complex geometries is used due to its link-wise interaction with the geometries. In addition it has a high scalability on large scale HPC systems due to its simple two step algorithm (Stream and Collide) and implicit nature.

The collision step is cell local including the computation while the streaming step requires neighbor information. Since the LBM is suited only for cubic elements, a well studied octree data structure is employed to approximate complex geometries. In our APES framework almost perfect load balancing is achieved by space-filling curve. MPI parallelism is used to communicate between cores. In this work, we present the performance of our lattice Boltzmann solver Musubi in APES framework with complex geometry. Two different applications: medical and electrodialysis are presented which requires simulation with complex geometry.

The remainder of this publication is structured as follows: At first the lattice Boltzmann method (LBM) is introduced (Sect. 2) and its implementation called Musubi in the end-to-end parallel simulation framework APES is described (Sect. 3). Based on this two possible applications on highly complex geometries are presented in Sect. 4. On the one hand simulation results of bloodflow in patient specific aneurysm geometries including a simple clotting model for simulating the thrombus formation are shown in Sect. 4.1. On the other hand the flow through a spacer filled flow channel from electrodialysis process for sea water desalination is presented in Sect. 4.2. The HPC aspects including the performance and scalability of Musubi simulating flow in highly complex geometries are shown in Sect. 5.

2 The Lattice Boltzmann Method

The lattice Boltzmann method can be viewed as a special discretization of the Boltzmann equation [4, 20]. It describes the time evolution of a particle distribution function (*pdf*) f on a Cartesian grid. The discretized LBM is an evolution equation for the pdf's f_i along different directions defined on a discrete set of velocities $\boldsymbol{e_i}$. The LBM equation using the Bhatnagar Gross Krook (BGK, [1]) approximation for the collision term reads

$$\underbrace{f_i(\boldsymbol{x} + \boldsymbol{e_i}\delta t, t + \delta t) = \underbrace{f_i(\boldsymbol{x}, t) - \omega(f_i(\boldsymbol{x}, t) - f_i^{eq}(\boldsymbol{x}, t))}_{\text{Collision}}}_{\text{Streaming}} \tag{1}$$

where ω is the relaxation parameter with that the collision part relaxes towards the thermodynamic equilibrium f_i^{eq}. A common choice for the discretization is the D3Q19 model with 19 discrete velocities $\boldsymbol{e_i}$ in 3 dimensions. The equilibrium pdf for the D3Q19 model reads

$$f_i^{eq}(\boldsymbol{x},t) = w_i \rho \left(1 + \frac{\boldsymbol{e_i u}}{c_s^2} - \frac{\boldsymbol{u}^2}{2c_s^2} + \frac{(\boldsymbol{e_i u})^2}{2c_s^4}\right) \tag{2}$$

where ρ defines the density, $\boldsymbol{u}$ the current velocity, c_s the speed of sound and w_i different weights for the 19 discrete velocities $\boldsymbol{e_i}$. The current velocity $\boldsymbol{u}$ and density ρ can be calculated locally from the pdf's by the following equations

$$\rho = \sum_{i=1}^{n} f_i(\boldsymbol{x},t) \tag{3}$$

$$\boldsymbol{u} = \frac{1}{\rho}\sum_{i=1}^{n} f_i(\boldsymbol{x},t)\boldsymbol{e_i}. \tag{4}$$

The LBM algorithm can be decomposed into a streaming and a collision step (see Eq. 1). The collision is an element-local operation colliding the particle distributions and converging them towards f_i^{eq} while the streaming step is related to the free flight of particles and copying of links from adjacent elements to the local ones.

The Boltzmann equation can be related to the Navier-Stokes equation by choosing two different scalings for the asymptotic expansion. The incompressible equations can be developed by using the diffusive scaling [14] while the acoustic scaling leads to the isothermal, compressible equations for small density variations [5]

The WSS can be calculated element-local using the formulation presented by Krüger et al. [15].

3 Musubi as Part of the APES Framework

The Adaptable Poly-Engineering Simulator (*APES*) is a framework for simulating different physics and numerics [17]. The central part of the APES framework is a central octree based elemental mesh data structure (*TreELM*) provided as a library to the different components of the APES suite (see Fig.1). Each element in the octree is identified uniquely with a *treeID* holding all information about its position, level in the octree and by this its parent and children as well as its neighbors treeIDs. Besides the grid information the TreElm library acts as a central link between the mesh generator *Seeder*, the solver components and the post-processor *Harvester* (see Fig.2). It provides functionalities like writing of data files which can be used as restart files or to visualize simulation results, reading configuration files in Lua format [10] and a generic load balancing.

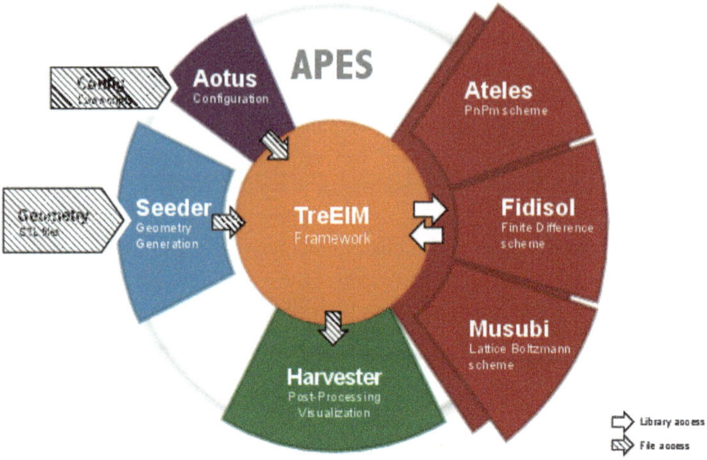

Fig. 1 APES schematic overview

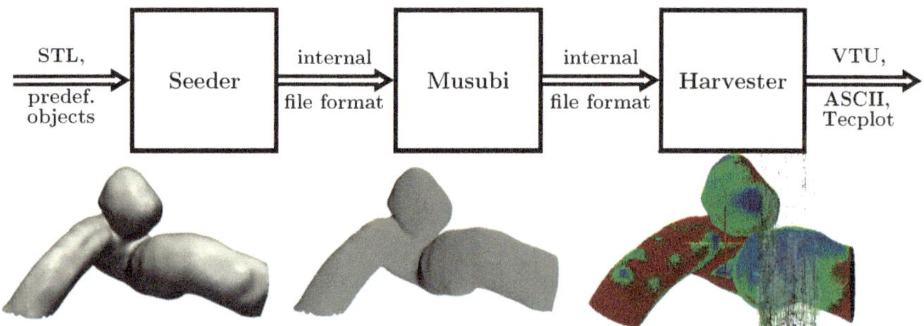

Fig. 2 The workflow in the APES framework

The octree is generated by the Seeder from either predefined spatial objects (e.g. rectangles, spheres, tubes, etc.) or geometries in STL-format (STereoLithography). The Seeder is capable of creating meshes with 2^{20} elements per dimension. After voxelizing the domain the Seeder uses a flooding algorithm to identify the simulation domain from the overall octree and the different types of boundaries are set according to the settings in the configuration file. To make the created grid available to the solvers in the APES framework the fluid elements and their additional information is dumped to disc in an efficient parallel way which can be loaded by the solvers in APES. This ensures that meshes can be used multiple times and do not have to be generated on the fly.

The solver component *Musubi* is based on the lattice Boltzmann method described in Sect. 2 which is known to give very good parallel performance results using several hundred-thousand cores (see [8, 21] and Sect. 5). It includes several computing kernels for 2D and 3D simulations using a list-based sparse matrix representation [19] of the fluid elements in the octree making the kernels vectorizable and highly efficient. Besides the kernels for pure fluid flow, Musubi offers several additional physical models for simulating incompressible fluids, multicomponent flow and passive scalar transport. To overcome the problems resulting from cubical elements especially in curved boundaries several higher order boundary conditions for solid [2] and open boundaries like pressure anti-bounce-back [6], do-nothing outflow [13], extrapolation [13] and non-reflecting boundaries [11] are provided.

In Musubi simulations using local grid refinement techniques are possible decreasing the number of elements and by this the workload on complex geometries where extremely small structures (e.g. stents, porous media or spacer) can be spatially resolved in rather huge simulation domains.

The post-processing tool of the APES framework is the Harvester. It can read data files written by the solver components as well as by the Seeder. It is capable of spatial- and temporal-reductions of the data, post-simulation calculation of physical quantities and reduction of physical quantities (norms, differences). The resulting data can be outputed in various common dataformats like VTU or Tecplot.

4 Applications on Complex Geometries

In this section two different numerical simulations on highly complex geometries are presented. In Sect. 4.1 the blood flow through a stented cerebral aneurysm including the formation of blood clots in the aneurysm bulge using a simple clotting model is presented. In Sect. 4.2 the spacer filled flow channel from electrodialysis process for sea water desalination is presented with the simulation setup used to perform scaling analysis of Musubi.

4.1 Thrombus Formation in Cerebral Aneurysms

Cerebral aneurysms are saccular extensions of blood vessels in the human brain. In case of rupture blood leaks from the vessel into the surrounding tissue causing irreversible damage or even death. Autopsy studies lead to the conclusion that 1–5 % of the adult population suffer from cerebral aneurysms [3] of which 1 out of 10,000 rupture [18]. Methods to treat cerebral aneurysms are *clipping*, *coiling* and the use of *stents*. In the clipping method a metal clip is placed at the aneurysm neck to stop the blood flow into the aneurysm and thus prevents it from rupturing. The main disadvantage of clipping is that the clip has to be placed using surgical techniques. Recent developments suggest that endonasal clipping approaches might be possible but these are only possible for aneurysms located in the path of the nasal cavity. Another approach is coiling where a metal coil is inserted through the femoral artery into the aneurysm using a catheter. In case of aneurysms with large aspect ratios stents are used to keep the coil inside the aneurysm. Besides these two 'classical' methods the use of flow diverter stents is becoming more and more popular. Stents are mesh like tubes which are inserted into the blood vessel to cover the neck of the aneurysm to change the blood flow properties initiating natural thrombosis in the aneurysm bulge. Currently the clinicians intuition and experience is the only basis on choosing the appropriate treatment strategy, once a cerebral aneurysm has been detected.

Ideally, multiscale numerical simulations on post-treatment flow properties and clotting dynamics in the aneurysm bulge might support the choice of suitable treatment strategies in general and flow diverter stents in detail. Due to the high complexity of the aneurysm and especially the stent geometry blood flow is simulated using the lattice Boltzmann method, which is proved to be optimal for large numerical simulations [21]. In addition to the complexity of the geometries the processes leading to clotting act on a vast set of temporal (μs to weeks) and spatial scales (Å to dm) making the use of simplification techniques like amplification [9] necessary.

To simulate the clotting process a simplification of the model proposed by Ouared et al. [16] and Harrison et al. [7] is used, taking the *wall shear stress (WSS)* as clotting initiator into account.

4.1.1 A Model for Thrombus Formation

To resolve the aneurysm with sufficient accuracy the mesh size and such the time step has to be very small. With a time step in the order of 10^{-5} s, more than eight billion iterations have to be performed for simulating a single day in real time. As mentioned before the upper temporal bound for the clot formation is in the order of weeks. This leads to the conclusion that simulating the whole process would be far too expensive even on the largest supercomputers.

To overcome these problems the technique of amplification [9] can be used. The main idea is to amplify the effect of the long-term process of thrombus formation and only simulate one or two cardiac cycles for the short-term process of blood flow. In the present model this can be achieved by e.g. increasing the probability p of an element to solidify or the growth rate.

As mentioned before the main condition used in the clotting model is the wall shear stress as an upper threshold. To prevent clots from developing in non-physical areas like near the outlet and inside the flow itself the *proximity* [7, 16] condition is used. This condition allows elements to solidify only in near wall or already solidified elements, preventing non-physical insulated clot formation in the bulk of the flow. In case the WSS condition and near-wall condition are fulfilled for a computational fluid element, it will turn into a solid element with a given probability p (see Fig.3). After the solidification step the simulation has to

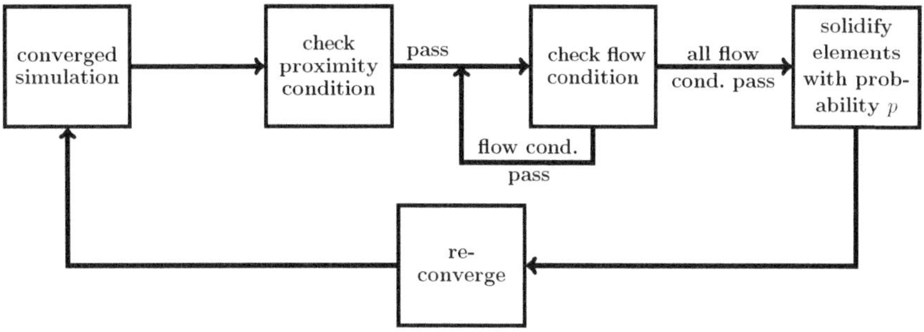

Fig. 3 The workflow for solidification of fluid elements

re-converge before starting with the next solidification step. This can be handled by an interval or a convergence criterion.

4.1.2 Simulation Setup

Based on the WSS condition mentioned in Sect. 4.1.1 a blood clotting simulation in a patient specific stented aneurysm has been performed. The grid size was chosen to be $\delta x = 6.5 \cdot 10^{-5}$ m resulting in approx. 45 million fluid elements, the time step to be $\delta t = 2.89 \cdot 10^{-5}$ s the inlet velocity to be $u_{in} = 9.435 \cdot 10^{-2}$ m/s. Using the inlet diameter as the characteristic length and the mean velocity $u_{mean} = u_{in}/2$ as the characteristic velocity resulting in a Reynolds number of 50. The density of blood was set to $\rho_{blood} = 1025\,\text{kg/m}^3$ and the kinematic viscosity to be $\nu_{blood} = 3.8 \cdot 10^{-6}\,\text{m}^2/\text{s}$.

The simulations have been performed on the superMUC at the LRZ-Munich for approx. 4 h on a full island (8,192 cores) using the Musubi lattice Boltzmann solver. As mentioned in Sect. 4.1.1 the simulation had to be amplified to reduce the number of iterations. Therefore the probability for an element was chosen to be constant $p = 1$ and the growth rate was amplified by letting only whole elements solidify. After the convergence of the flow the clotting model was switched on with a wall shear stress threshold of $1.037 \cdot 10^{-2}$ Pa.

4.1.3 Simulation Results

Figure 4 shows the thrombus formation in the aneurysm bulge in terms of solidified fluid elements (marked in red) for the non-stented (Fig. 4a) and stented (Fig. 4b) case. As can be seen the amount of solidified fluid elements increases significantly inside the aneurysm bulge after deployment of the stent. The solidified elements in the vessel itself are due to gaps between the inserted stent and the vessel wall. The growth process stopped in these areas as soon as the gaps were filled.

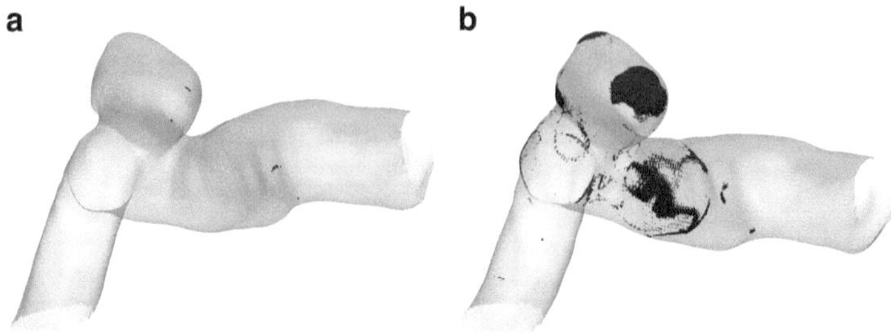

Fig. 4 Differences in the thrombus formation for the non-stented (**a**) and stented (**b**) aneurysm

4.2 Spacer Filled Flow Channel in Electrodialysis

There is growing demand to supply mankind with drinkable water. Sea water desalination process is one of the solution to this problem. Among various desalination processes, electrodialysis is one of the efficient process which uses ion exchange membranes to separate salt ions from sea water under the influence of an applied electrical field. This process contains two selective permeable membranes named as cation exchange membranes (CEM) and anion exchange membranes (AEM) arranged periodically. When electrical field is applied, ions start separate from salt water and CEM and AEM allows only cations and anions to pass through respectively. This results in alternative dilute channels and concentrate channels. Finally, the desalinated water is collected from dilute channels.

In electrodialysis process, ion exchange membranes are separated by complex structure named as spacer. This spacer structure act as a mechanical stabilizer. The design configuration of this spacer structure has a influence on the pressure drop in the channel which in turn affects the power consumption of electrodialysis module. It also influence the transport of ions through the membranes. Due to these facts, the optimization of spacer geometry play a vital role in the development of electrodialysis module. Figure 5 shows the structure of laboratory scale interwoven spacer geometry which is used in our simulation and scaling analysis.

4.2.1 Simulation Setup

In this paper, the spacer filled flow channel shown in Fig. 6 is used as a simulation setup. In the figure, L is the length of the channel, W is the width of the channel and H is the height of the channel. The height of the channel is four times the radius of the spacer filament. Figure 6a also shows that the flow is along the length L in x-direction and the channel is periodic along the width W in z-direction. The channel walls at $y = 0$ and $y = H$ are treated as solid walls. In the laboratory scale, the extent of the spacer sheet is $20 \times 10 \times 0.04$ cm. The channel with a single spacer element as in Fig. 6 has a dimension $0.2 \times 0.2 \times 0.04$ cm.

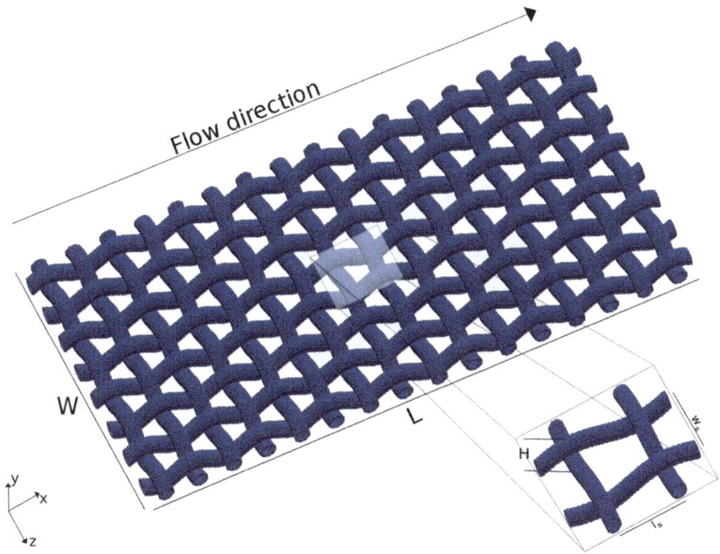

Fig. 5 Structure of a spacer utilized to stabilize the fluid channels of the spacer stack mechanically. The structure of this spacer has a significant impact on the total energy consumption of the stack

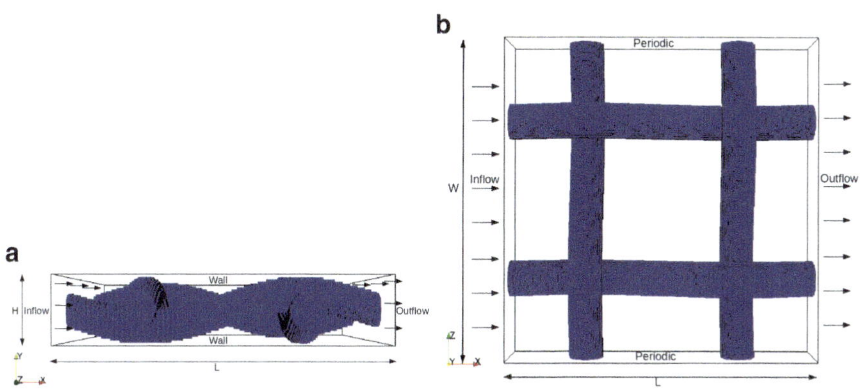

Fig. 6 Schematic layout of the simulation setup with interwoven spacer geometry. (**a**) Side view of the channel. (**b**) Top view of the channel

The distance between two spacer filaments is 0.1 cm. For scaling analysis, the number of spacer elements along the length is increased by increasing the length of the channel.

The following boundary conditions are used in this test case: The solid walls at $y = 0$ and $y = H$ are treated as simple bounce back boundaries resulting no-slip on walls. At inlet $x = 0$, the velocity bounce back boundary is applied [11] and at outlet $x = L$, extrapolation outflow boundary condition from Junk et al [12] is applied.

5 Scalability and Parallel Efficiency of Fluid Flows in Complex Geometries

This section describes the performance of Musubi with complex spacer geometry on the Cray XE6 system Hermit at HLRS. The Hermit system provides of 3,552 computing nodes with AMD Interlagos on two sockets where each socket has 16 cores resulting in 32 cores per node. For our performance analysis, up to 1,024 computing nodes or 32,768 cores are used. Only MPI parallelism is considered here. Both, intranode and internode performance are measured i.e performance within a single node (up to 32 cores with as many MPI processes) and between multiple nodes. For scaling analysis, the problem size is increased from a single spacer element length of 0.2 cm (Zoomed part in Fig. 5) with 66,000 elements to the laboratory scale spacer length of 20 cm with 66 million elements. In the width of this channel slice, a periodic boundary is assumed.

In Lattice-Boltzmann codes the measurement of the lattice updates per second is commonly used to compare the performance. This number of lattice updates per second (LUPs) will be used in the following part of the paper, as it provides a direct estimation on how long a given simulation, that requires a certain number of lattice nodes and time step updates will take. We represent the behavior of the code in terms of performance per execution unit, that is per node or per core, to get a clearer impression of the performance independent of the number of used execution units. An ideal parallel execution is expected to just replicate the serial behavior on each execution unit. However, the execution performance is influenced by cache usage, non-computational implementation overheads, vector lengths, communication times and so on.

In our analysis we used LBM model with $D3Q19$ layout with the BGK collision operator. The optimized compute kernel was used and it requires only 150 floating point operations per element. In our LBM solver, instead of communicating all probability density functions on the communication surface between the processors, only the required links of probability density functions are communicated. This reduces the MPI buffer size and bandwidth driven communication times.

Figure 7a shows the performance per core over the problem size per core for various number of cores within a single node. We cannot see the cache region in this figure since smallest problem size i.e. single spacer element with 66,000 elements is above cache limit. Performance almost flattens out but with increase in problem size the performance drops for smaller number of cores due to frequent access to slower memory. However, the usage of full node (32 cores) gives better performance. We achieved a sustained performance of roughly 4.2 % on a full node and the Hermit system has a theoretical peak performance of 294.4 GFLOPS per node.

The internode performance map is shown in Fig. 7b with problem size per node in horizontal axis and performance per node in vertical axis for various total number of nodes. The performance map combines both, weak scaling and strong scaling. Weak scaling can be measured by the vertical comparison of points between different lines

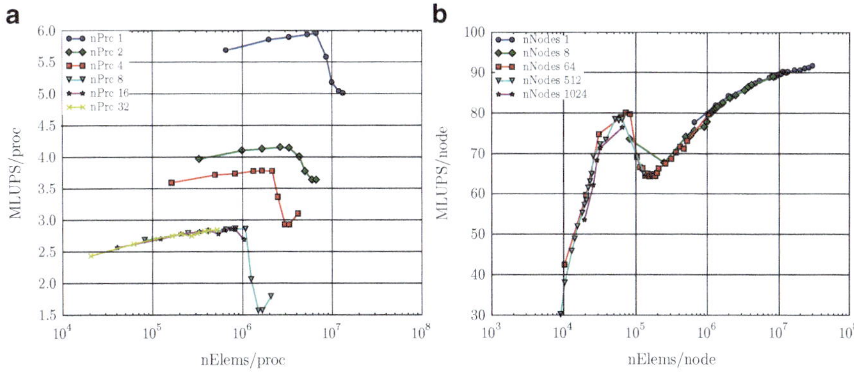

Fig. 7 Intranode (**a**) and Internode (**b**) performance map with spacer structure

for the node count i.e. fixing the number of elements per node. The closer the points are located to each other the better the weak scaling. Strong scaling on the other hand is not as easily seen in the performance map, but can be derived by moving to the left when increasing the number of nodes. This reduces the number of elements per process with increasing node counts, as required by strong scaling with a fixed overall problem size.

In our solver framework, almost perfect computational load balancing is achieved with the help of space-filling curve i.e there is at most single element difference between the partitions. However, due to the irregular domain and the large number of walls in spacer filled channel, the communication surface between different processes might vary drastically, resulting in a large imbalance of communication costs. The effect of this load imbalance can be noticed in the performance map shown in Fig. 7b with a relatively high performance drop for smaller domains per node.

The steep slope after the cache effect in both models Fig. 7b is supposedly due to load imbalances in the communication times, caused by the walls of the spacer geometry. This can be explained with the help of periodic testcase performance Fig. 8, where the performance flattens out after the cache effect with a small slope. The main difference between the two simulations is the exactly balanced communication effort in the one case and a less than optimal balancing for the complex spacer geometry. This can be resolved using a dynamic load balancing algorithm to distribute the simulation domain on each processor at runtime according to the actual load.

From Fig. 7b, it can be seen that weak scaling works fine on different process counts for all problems, that fit into memory down to the cache-sized problems, where the communication gets dominant and the performance per node drops down with a steeper slope.

An explicit plot for weak scaling for two different problem size per node is shown in Fig. 9a. One with approximately 21,000 elements per node which is below cache

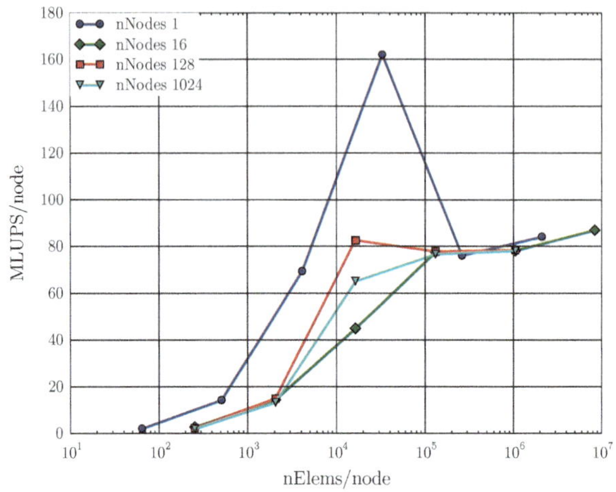

Fig. 8 Internode performance map for periodic cubic simulation domain

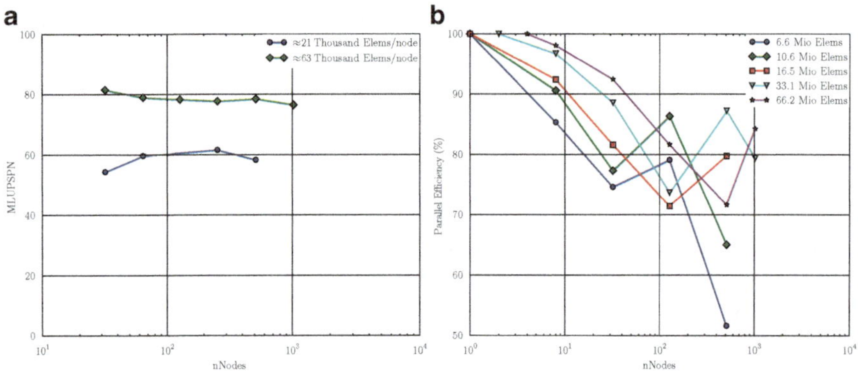

Fig. 9 (**a**) Weak scaling for two different number of elements per node with spacer structure for different number of nNodes. Ideal weak scaling is a straight line. (**b**) Strong scaling parallel efficiency for different problem sizes with spacer structure

and other 63,000 elements per node which almost fit in the cache. In both cases, weak scaling is almost perfect with only a small drop in the performance for larger counts of compute nodes.

A dedicated graph for strong scaling is shown in Fig. 9b with number of nodes on the horizontal axis and parallel efficiency (%) in the vertical axis. Here, testcases with problem sizes of roughly 6.6, 10.6, 16.5 and 66.2 million elements are used. Here, with a problem size of 6.6 million elements, the performance drops from 1 to 1,024 nodes because the communication dominates the computation. The peak in the plots defines the problems which fits in the cache. The largest problem size of full spacer length with 66.2 Million elements fits into cache for 1,024 nodes. This problem requires minimum of four nodes to fit in the memory. Thus our solver can scale up to full spacer length with efficient usage of 1,024 compute nodes of the Hermit.

6 Conclusion and Outlook

In this paper, we presented our APES framework and the lattice Boltzmann solver named MUSUBI. We introduced two different applications: thrombus formation and electrodialysis. Both applications involve fluid flow simulations with complex geometry. We presented simulation results of clot formation and scaling results with spacer geometry. The scaling analysis shows that our code has good scalability up to thousands of nodes with irregular complex geometry. In the future, we will investigate on the dynamic load balancing to reduce communication time.

References

1. P L Bhatnagar, EP Gross, and M. Krook. A Model Collision Processes in Gases. I. Small Amplitude Processes in Charged and Neutral One-Component Systems. *Phys. Rev*, 94: 511–525, 1954.
2. M Bouzidi, M Firdaouss, and P Lallemand. Momentum transfer of a Boltzmann-lattice fluid with boundaries. *Physics of Fluids*, 13(11):3452–3459, 2001.
3. J L Brisman, J K Song, and D W Newell. Cerebral aneurysms. *New England Journal of Medicine*, 355(9):928–939, 2006.
4. C Cercignani. *Theory and application of the Boltzmann equation*. Elsevier, 1976.
5. S Chen and G.D. Doolen. Lattice Boltzmann method for fluid flows. *Annual review of fluid mechanics*, 30(1):329–364, 1998.
6. Irina Ginzburg, Frederik Verhaeghe, and Dominique d'Humieres. Two-relaxation-time Lattice Boltzmann scheme: About parametrization, velocity, pressure and mixed boundary conditions. *Communications in Computational Physics*, 3(2):427–478, 2008.
7. S E Harrison, Smith S M, J. Bernsdorf, D R Hose, and P V Lawford. Application and validation of the lattice Boltzmann method for modelling flow-related clotting. *Journal of biomechanics*, 40(13):3023–3028, January 2007.
8. Manuel Hasert, Kannan Masilamani, Simon Zimny, Harald Klimach, Jiaxing Qi, Jörg Bernsdorf, and Sabine Roller. Complex Fluid Simulations with the Parallel Tree-based Lattice Boltzmann Solver Musubi. *Journal of Computational Science*, pages 1–20.
9. A G Hoekstra, A Caiazzo, E Lorenz, J-L Falcone, and B Chopard. Complex automata: Multi-scale modeling with couples cellular automata. In A G Hoekstra, J Kroc, and P M A Sloot, editors, *Simulating Complex Systems by Cellular Automata*, Understanding Complex Systems, chapter 3, pages 29–57. Springer, 2010.
10. Roberto Ierusalimschy, Luiz Henrique de Figueiredo, and Waldemar Celes Filho. Lua-an extensible extension language. 1995.
11. Salvador Izquierdo and Norberto Fueyo. Characteristic nonreflecting boundary conditions for open boundaries in lattice Boltzmann methods. *Physical Review E*, 78(4), October 2008.
12. M Junk and Z Yang. Outflow boundary conditions for the lattice Boltzmann method. *Progress in Computational Fluid Dynamics*, 8:38–48, 2008.

13. M Junk and Z Yang. Pressure boundary condition for the lattice Boltzmann method. *Computers and Mathematics with Applications*, 58(5):922–929, September 2009.
14. Michael Junk, Axel Klar, and Li-Shi Luo. Asymptotic analysis of the lattice Boltzmann equation. *Journal of Computational Physics*, 210(2):676–704, December 2005.
15. T Krüger and F Varnik. Shear stress in lattice Boltzmann simulations. *Physical Review E*, 2009.
16. R Ouared and Bastien Chopard. Lattice Boltzmann simulations of blood flow: non-Newtonian rheology and clotting processes. *Journal of Statistical Physics*, 121(1):209–221, 2005.
17. Sabine Roller, Jörg Bernsdorf, Harald Klimach, Manuel Hasert, Daniel Harlacher, Metin Cakircali, Simon Zimny, Kannan Masilamani, Laura Didinger, and Jens Zudrop. An Adaptable Simulation Framework Based on a Linearized Octree. In *High Performance Computing on Vector Systems 2011*, pages 93–105. Springer, 2012.
18. Wouter I Schievink. Intracranial aneurysms. *New England Journal of Medicine*, 336(1):28–40, 1997.
19. M Schulz, M Krafczyk, J. Tölke, and E. Rank. Parallelization strategies and efficiency of CFD computations in complex geometries using Lattice Boltzmann methods on high-performance computers. pages 115–122, 2002.
20. S Succi. *The lattice Boltzmann equation for fluid dynamics and beyond.* Oxford University Press, USA, May 2001.
21. J Zudrop, H Klimach, M Hasert, K Masilamani, and S Roller. A fully distributed CFD framework for massively parallel systems. *cug.org*, April 2012.

IMD: A Typical Massively Parallel Molecular Dynamics Code for Classical Simulations – Structure, Applications, Latest Developments

Johannes Roth

Abstract We give a short description of IMD, a classical molecular dynamics package for the simulation of condensed matter. The properties of molecular dynamics simulations will be given with examples of their implementation in `IMD`. We further report on multi-scale simulations with `IMD`, the determination of accurate interactions with `potfit` and the porting of `IMD` to GPUs.

1 Introduction

Molecular dynamics (MD) simulations on the atomistic level play an important role today in science and industry. It can be applied – in different ways certainly – from the statistical mechanics of sub-atomic particles up to the most accurate calculation of the orbits of planets or the rings of Saturn. Here, however, we will concentrate on molecular dynamics simulations in condensed matter physics. We start with an explanation of the components of a molecular dynamics simulations and used that for the introduction of the implementation of MD in the **I**TAP **M**olecular **D**ynamics package `IMD`. For more details see the basic publications [16, 22] and the webpage imd.itap.physik.uni-stuttgart.de/userguide/imd.html. We will then elaborate a little about one of the most important parts, namely of the modeling of the interactions. Since we want to apply MD to atoms where the interaction is mediated by electrons through metallic, covalent, of ionic bonding, one should solve the problem by quantum mechanics. The drawback is that this would limit the size of the simulations from several hundred up to a few thousand atoms, or to some hundred thousand particles if we use tight-binding methods. Our goal, however, is to simulate big systems beyond millions of particles for up to μs to be able do deal

J. Roth (✉)
Institut für Funktionelle Materalien und Quantentechnologien, Universität Stuttgart, Stuttgart, Germany
e-mail: johannes@itap.physik.uni-stuttgart.de

M.M. Resch et al. (eds.), *Sustained Simulation Performance 2013*,
DOI 10.1007/978-3-319-01439-5_5,

with long-range effects. We do not want to discuss the applicability of classical MD simulations in detail but refer the reader to the Bohr-Oppenheimer approximation. If applicable, we still have to model the interactions of our sample which can be done with *potfit*, a program that fits the forces, energies and stresses of the classical interactions to quantum mechanical simulations of a database of small samples.

The paper is organized as follows: we start with classical molecular dynamics simulations, then we introduce `potfit`. We will further give a number of recent examples of the application of `IMD`. Then we will give a short account on parallelization and benchmarking `IMD`. We will continue with the attempts to port `IMD` to GPUs. We will end with a comment on very big simulations.

2 Classical Molecular Dynamics Simulations

The basic idea of classical molecular dynamics simulations is simple: if there are N particles, integrate the $3N$ Newtonian second order equations of motions or equivalently the $6N$ Hamiltonian first order system of equations of motion. Since this is an initial value problem of ordinary differential equations, we have to specify in the case of point-like particles without internal structure $6N$ starting values which are $3N$ coordinates and $3N$ velocities or momenta.

2.1 The Molecular Dynamics Steps

In the next sections we will describe the components of a MD simulation. Details of the different steps can be found in the basic book by Allen and Tildesley [2]. Some parts like storing data on tapes are quite outdated, but the major ideas presented in the book are still valid. The examples and features described are those available in `IMD` [16, 22].

1. **First step: Initial conditions**
   The coordinates are typically given by the atom positions in the sample under study. If it is a fluid or gas, the coordinates might be generated by a random number generator. For a crystal, the coordinates are found in crystallographic databases or may be determined from real experiments. Often the coordinates are obtained from model systems for which structural properties or phase diagrams are to be studied.
   Normally, the velocities or momenta of the particles are not known. Thus they are typically generated with a random number generator. Since random number generators produce homogeneous distributions of random numbers, they have to be converted to yield the desired Maxwell-Boltzmann distribution (see for example [2]). At the same time they are scaled to give the desired temperature via the equipartition theorem

$$2E_{\text{kin}} = 3Nk_{\text{B}}T$$

where E_{kin} is the kinetic energy, N the number of particles, k_B is the Boltzmann factor and T is the instantaneous temperature.

2. **Second step: Interactions**
   For a long time, when computers were expensive and small, these were model interactions like the Lennard-Jones pair potential. Such interactions are still applied in statistical mechanics were people are interested in generic features of interactions. If properties of materials should be studied with high accuracy, the methods of choice is for example force-matching. Ab-initio calculations are carried out to obtain forces which are then fitted to obtain a classical interaction for the MD simulation. This can be done with *potfit* for example and will be explained in more detail in Sect. 3. In organic chemistry and soft matter physics especially standardized two-, three-, and four-body interactions called force-fields exist for modeling bond lengths, angles, and dihedral angles.
3. **Third step: Boundary conditions**
   The surface to volume ratio is important even for the largest simulations in three dimensions. Thus one has to specify boundary conditions. If surface effects are negligible, open or free boundaries are applied, which effectively means that a cluster of particles is simulated. If surface effects should be avoided, periodic boundary conditions are applied, which means theat the left and right, upper and lower and top and bottom surfaces are pairwise identified and if a particle leaves the box on one side it will enter on the opposite side again. Thus there is no boundary at all and the simulation box is a torus. In solid state physics and materials sciences it is often of interest to deform the sample by straining it for example. This can be done by fixed boundary conditions, where the atoms at the surface are moved according to a given trajectory or an applied forces. Finally, there are more advanced boundary conditions, for example the Lees-Edwards boundary conditions [2] which allow to imprint a flow on the sample without applying forces.
4. **Fourth step: Integrators and linked lists**
   While there are basically two formulations of the standard equations of motion for classical particles in a micro-canonical ensemble with constant NVE (N number of particles, V volume, E total energy), namely the Newtonian and Hamiltonian, there is an arbitrary number of integrators. The Hamiltonian equations of motion are

$$\frac{d\mathbf{p}_i}{dt} = -\nabla V(\mathbf{r}_1, \ldots, \mathbf{r}_N)$$

$$\frac{d\mathbf{r}_i}{dt} = \frac{\mathbf{p}_i}{m_i}$$

   with the interaction potential $V(\mathbf{r}_1, \ldots, \mathbf{r}_N)$ depending on all coordinates $\mathbf{r}_i$ and $\mathbf{p}_i$ the momenta of the N particles. Typically this $6N$ dimensional system of differential equations is solved by discretizing the time into steps small enough that the basic vibrational frequencies and the fastest motions can still be represented.

Today typical choices of the integrators are the modifications of the so-called Verlet- and leap-frog algorithms. Principally they are inferior to the older Gear-predictor-corrector algorithms or to Runga-Kutta-type integrators, but they have many advantages: they require less storage, which is especially useful for communication on massively parallel supercomputers. Furthermore, they are similar to so-called symplectic integrators which respect the inherent symmetry of the equations of motion and are thus more stable even if there nominal accuracy is lower than that of higher oder integrators. A typical leap-frog integration scheme looks like this:

$$\begin{aligned} \dot{\mathbf{p}}_\mathbf{i}(t+\Delta t/2) &= \dot{\mathbf{p}}_\mathbf{i}(t-\Delta t/2) + \mathbf{f}_i(t)\cdot\Delta t \\ \dot{\mathbf{r}}_\mathbf{i}(t+\Delta t) &= \dot{\mathbf{r}}_\mathbf{i}(t) + \frac{1}{m_i}\mathbf{p}_i(t+\Delta t/2)\cdot\Delta t \end{aligned}$$

It is called leap frog since the evaluation of positions and coordinates are shifted by half a time-step. First the force $\mathbf{f}_i(t)$ for each particle i is computed and added to the previous momenta $\dot{\mathbf{p}}_\mathbf{i}(t-\Delta t/2)$, then the positions $\dot{\mathbf{r}}_\mathbf{i}(t)$ are updated and the new forces $\mathbf{f}_i(t+\Delta t)$ will be calculated.
The most time consuming part of the simulations is the evaluation of the forces. If the potential $V(\mathbf{r}_1,\ldots,\mathbf{r}_N)$ is approximated in the most simple approach by pair potentials $V(\mathbf{r}_i,\mathbf{r}_j)$ which act between particle i and j, then the complexity of the algorithm is still $O(N^2)$ and uses 80 % of computation time or more. If the interactions can be made short-ranged by cutting them off at certain distance, then it is possible to introduced linked lists for the administration of the interactions (see [22] for details on `IMD`.) Typical examples are metals. Long-range interactions between ions or polar molecules like water have to be treated in a different way (see Sect. 3.1). For the linked lists the simulation box is divided into cells with the size of the interaction range. Each atom is assigned to a cell which can be done by local information only. Each atom can only interact with its neighbors in the same cell or in neighboring cells due to the cutoff. Together with actio equal to reactio this leads to 13 partner cells in three dimensions. Now the complexity of the algorithm is only $O(N)$, and the goal is to reduce the pre-factor, i.e. the number of particles in each cell. Especially in the case of three-body interactions $V(\mathbf{r}_i,\mathbf{r}_j,\mathbf{r}_k)$ and many-body interactions this is advisable: the cells are not used to compute directly to compute the forces but to determine the interaction partners. If a certain "skin" is added for particles that are close to interaction, then this list can be recycled and need not be updated at every time step. Such tables are called Verlet-tables [2] although they were originally introduced in a different way which had still an order of $O(N^2)$ complexity.

5. **Fifth step: Influence from the outside**
Up to now we have studied an isolated micro-canonical *NVE* ensemble. If constant temperature as in a canonical *NVT* ensemble or even constant pressure *NPT* should be simulated, then the integration scheme has to be modified. Further extensions are required if the sample is stressed, deformed, or if flow is applied. Shock waves [13–15] or laser irradiation [19, 20] are other influences which require special treatment.

The method of choice to simulate constant temperature is the Noseé-Hoover thermostat. It can be shown that it leads to a canonical ensemble in limit of infinitesimal time steps. First the instantaneous temperature $T(t)$ is determined by applying the equipartition principle

$$\frac{3}{2}Nk_{\mathrm{B}}T(t) = \sum_{i}^{N} \frac{\mathbf{p}_i^2}{2m_i}.$$

Then system is derived towards the desired temperature T by feedback applied to the forces:

$$\dot{\mathbf{p}}_{\mathbf{i}} = -\nabla_i V - \eta_T \mathbf{p}_i$$

where η is determined by a new differential equation

$$\dot{\eta} = \nu_T \left\{ \frac{T(t)}{T} - 1 \right\}.$$

The strength of the coupling is determined by the frequency ν_T which can be determined from the Einstein frequency and its relation to the average force (for details see [2]).
The simulation of constant pressure works in a similar way. Here we start from the pressure equation for the instantaneous hydrostatic pressure $P(t)$ of an interacting material:

$$P(t) = \rho k_{\mathrm{B}} T + W/V$$

with density $\rho = N/V$, volume V and virial

$$W = \frac{1}{3}\sum_{i}^{N} \mathbf{r}_i \cdot \mathbf{f}_i.$$

The desired pressure P is again achieved by feedback, not applied to positions and momenta

$$\dot{\mathbf{r}}_{\mathbf{i}} = \frac{\mathbf{r}_i}{m_i} + \xi \mathbf{r}_i$$

$$\dot{\mathbf{p}}_{\mathbf{i}} = -\nabla_i V - \xi_T \mathbf{p}_i.$$

The coupling parameter ξ_T is much more difficult to determine than η_i. In principle it should not alter the frequency spectrum of the simulation. In praxis a first good choice is to set xi_T equal to η_T. Then a few tests with a variation of the parameter by a magnitude up or down will show if the coupling an will work correctly. Although the coupling parameters are in principle temperature dependent, we observe that the same parameters can be used at least in the whole solid range of the phase diagram. The method can easily be extended to constant stress simulations of solids with uniaxial loading for example.

Intrinsic flow can be simulated by the Lees-Edwards boundary conditions. The equations of motions should also be modified as given here in two dimensions:

$$\begin{aligned} \dot{x}_i &= \frac{p_{xi}}{m_i} - \gamma y_i & \dot{y}_i &= \frac{p_{yi}}{m_i} \\ \dot{p}_{xi} &= -\nabla_{xi} V + m_i \gamma \dot{y}_i & \dot{p}_{yi} &= -\nabla_{yi} V \end{aligned}$$

The strength of the flow is determined by γ.

6. **Sixth step: Data evaluation**

   MD simulations will directly yield potential and kinetic and thus the total internal energy by summing over all particles. To compute the free energy is not so trivial since it is not a mechanical quantity depending on coordinates an momenta. Methods to compute the free energy are given for example in [2]. Further data which are directly available are total and local stresses and displacement fields. Elastic constants can be computed by systematically deforming the sample along the required modes. Transport coefficients can be obtained from equilibrium simulations via the Green-Kubo relations or by non-equilibrium simulations, for example by applying a temperature gradient. Correlation functions are obtained if the required data are added up during simulations. These and many other observables are implemented in `IMD` together with utilities to evaluate them. Diffraction patterns for example can be calculated from the correlation functions through fast Fourier transform.

7. **Seventh step: Visualization**

   Many three-dimensional processes that occur in the bulk and dynamical procedures can only be detected by modern visualization methods. There are many programs today which allow the direct rendering of atoms like `VMD` (www.ks.uiuc.edu/Research/vmd), `ovito` (www.ovito.org) or `MegaMol` (svn.vis.uni-stuttgart.de/trac/megamol) for example. Typically these programs allow to color code the particles according to selected observables. `IMD` allows to preselect the particles to be visualized, for example only those which are at the boundary or belong to a defect or form a cluster close to a crack surface.

   All the visualization programs also allow to produce movies from sequences of the particle configurations. The movies are especially helpful for the determination of dynamical processes.

Molecular dynamics simulations can be used to study equilibrium problems, like structure as a function of pressure and temperature or the stability of grain boundaries. But they can also be used to determine transport coefficients from fluctuations like diffusion constants or flip frequencies for example. Beyond that we run non-equilibrium simulations close to equilibrium, where the relaxation time is very short. This includes for example the study of plastic deformations, dislocation motion, crack propagation, shock waves or laser ablation, to name only a few processes were IMD has been used.

3 Realistic Interactions and *potfit*

If real properties of materials shall be reproduced with great accuracy it is unavoidable to use the best available interactions. Typically these are quantum mechanical ab-initio descriptions. But for mechanical studies for example samples with several million or hundred million atoms are required which are way beyond the size that can be simulated by ab-initio methods. Therefore a program was developed in close relation to IMD named potfit (potfit.sourceforge.net/wiki/doku.php) [5]. The idea is the following: compute forces, energies, stresses and so on by ab-initio calculations and fit classical interactions applicable to molecular dynamics simulations to match them.

The first step is the creation of a database of samples tractable with ab-initio calculations. This is a crucial step, since the members of the database have to represent the space of application intended. For example if cracks have to be simulated, then the database should contain samples with open surfaces, if high temperatures have to be simulated, then the database should contain samples from classical high temperature simulations, and so on.

The second step is the selection of the interaction. Currently available are simple pair interactions, embedded atom potentials and angular dependent potentials for metals and electrostatic interactions. These have the same format as in IMD. Further interactions can easily be added.

The third step is the choice of the interaction representation. You can choose between splines, where you have to specify the number and position of the supporting points together with the weight, or you can choose a functional form where then the parameters of the representation are fitted.

The last step is the choice of the fitting method: conjugate gradient methods and evolutionary algorithms are possible.

The data that can be fitted are forces, energies and stresses, and each variable can be weighted as desired.

The main problem with the free spline is that it requires data fora good fit where none are available: especially in crystals atoms are placed at discrete atom shells with large gaps in between. Even for heavy distortions these gaps cannot be sampled. Thus the fitted interaction may take a weird shape. But if it is used in simulations, then the particles will find these intermediate ranges which lead to strange results.

While force-matching works well for monatomic and binary samples it becomes less useful for ternary compounds where the fitting space gets too large or the functional shape of the interactions might not be representative enough of the true interactions.

3.1 Long-Range Interactions

As noted in Sect. 2.1 it is not possible to cutoff long-range Coulombic or dipolar interactions directly. The most accurate method is the Ewald summation. It cannot be used for large simulations since its complexity is $O(N^{3/2})$. Hierarchical methods, FFT and multipole methods reach $O(N \log(n))$ or close to $O(N)$, but they typically do not fit very well into our molecular dynamics scheme with domain decomposition for parallelization (for details see [16, 22]). Wolf et al. [26] have developed a method which is applicable to solids and liquids, where the charges are compensated and thus the interactions fall of fast enough. This method has been implemented in IMD [7] and applied successfully. If the cutoff complies to certain conditions as described in [7], then the Wolf summation is an $O(N)$ method with a cutoff more than twice the typical cutoff for short-range interactions, and thus rather costly. But it is still up to three orders of magnitude faster than the pure Ewald summation.

Subsequently we have extended the method to the simulation of induced dipoles as it is required for oxides. The model of Tangney and Scandalo [23] has been used for the description of the interaction. Again the implementation has been tested and applied for several studies of silica, alumina, and magnesia [3, 4, 7, 10]. The original parameters which where used by Tangney and Scandalo together with the Ewald summation lead to good results also with the Wolf summation, but there was still an improvement after re-fitting the parameters directly for the Wolf summation.

For the Tangney and Scandalo model the static polarization and the charges of the ions are given from literature or from force-matching and are thus constant during simulation. This is no longer possible if interfaces between metals and metal oxides shall be simulated since the atoms in the metal are neutral while ionized in the oxide, and the charge will vary across the interface. Streitz and Mintmire proposed a model [21] that can deal with this situation. The charges become variable and are determined by minimization of the chemical potential. While Streitz and Mintmire compute the chemical potential by inverting a huge matrix which is rather time consuming and not applicable to large systems, we have implemented a conjugate gradient algorithm which is much better since we have to find a parabolic minimum.

The problem with the chemical potential is that it requires a minimization of the whole system which is very costly on parallel computers. Thus we will have to use external libraries to tackle this problem. Since the chemical potential will vary only close to the interface it might be possible to find a more local treatment as for example in [9] for impurities.

4 Some Examples of Recent Simulations

IMD has been applied to many classical problems of materials sciences and solid state physics. We will not recall them here. We will rather present some applications where IMD has been used in a multi-scale environment. One of the older examples was a simulation by Büehler et al. [8], where the very complicated and time-consuming ReaxFF interactions have been calculated on the fly in an molecular dynamics simulation with IMD. Another application was the determination of the dynamical structure factor for Mg_2Zn_{11} intermetallics [11]. A long simulation of the material was run with IMD to produce a trajectory as input of the nMoldyn program (dirac.cnrs-orleans.fr/plone/software/nmoldyn) which could then be applied to obtain the phonon dispersion relations.

Another case was even more complicated. $CaCd_6$ is a material built of clusters including tetrahedra which show a phase transition from oriented to rotating. First an accurate interaction was determined by ab-initio calculations and force matching. Then classical simulations were run with IMD to obtain the energy differences between different correlated orientations of neighboring tetrahedra. Since MD was still too slow to study the phase transition directly, the data where used to set up a Monte-Carlo simulation with the transition energies obtained from the MD simulations. With Monte-Carlo the behavior of the system could be determine as a function of temperature and the phase transition could be observed at a temperature which is compatible with experimental observations [6].

As a last example we want to mention laser ablation simulations. Here the problem is that the interaction with the laser and the heat conduction is govern by electronic processes which cannot be described by classical molecular dynamics. But the fast heat conduction requires huge samples on the other hand. There is a two-temperature continuum model available with separate temperature for the electrons and the lattice or atom cores. This model works rather well if solved by finite difference methods and can predict many properties of laser ablation. However, it cannot give information on the atomistic level like the creation of defects, the production of gas and drops, and so on. We have combined the two-temperature model with MD, thus solving the electronic part with finite differences and the atomistic part by molecular dynamics simulations [19, 20]. We have been able to predict melting depths and ablation thresholds. For the simulation of the ablation plume composition, however, the electronic heat conductivity is still to fast and has to be switched of. Nonetheless, the results compare rather well to experiments.

5 Parallelization

IMD is parallelized with MPI and domain decomposition as described in [16, 22]. While this scheme works very well for samples with periodic boundary conditions it has some drawbacks for samples with open surfaces like crack propagation, indentation, shock waves or laser ablation. A more dynamical load balancing scheme has not been attempted to be implemented since the load distribution depends strongly on the kind of simulations. Meanwhile an OpenMP parallelization level has been added especially for cases where MPI does not work as for Ewald summation or certain correlation functions in conjunction with Fourier transforms. It turns out that OpenMP leads to increasing performance only up to four to eight compute cores, depending on the machine. A serious problem is that the compute cores may change there memory location from thread to thread, which could be avoided by setting environment variables.

In conclusion we can state that the MPI version yield always the highest performance, irrespective of shared or distributed memory. We found no case where OpenMP lead to a big improvement of performance.

6 Benchmarking IMD

IMD was written especially for massively parallel computers. Benchmarks have been published in [16,22]. Although the numbers have changed, the overall behavior is still valid [17,18]. This is a nearly perfect weak scaling and a good strong scaling. Recently this has been confirmed by Cray who tested IMD on the then Top500 machine Jaguar with up to 1.3×10^{11} particles and on the HLRS Hermite. If about a million particles are available per compute unit, then the degradation of performance is about 10 % up to the full machine. IMD could reach a considerable amount of the peak performance of the machines without special adaption or libraries.

7 Porting IMD to GPUs

In recent years it has been realized that graphics processing units (GPUs) yield a huge computing capacity after the advent of CUDA, a language which allows to program GPUs efficiently, the topic has gained speed. Nvidia, one of the producers of graphics cards and the creator of CUDA promises speedups up to 100-fold. In reality the speedups are much lower, often they are much below 10. The are MD programs that are written especially for GPUs, like HALMD (halmd.org) or HOOMD (codeblue.umich.edu/hoomd-blue) which are very fast and take heavy advantage of GPUs, but they are rather specialized for certain purposes. On the other hand, there are general purpose programs which are very powerful like lammps (http://lammps.sandia.gov/), GROMACS (www.gromacs.org/) or IMD which are difficult to port to GPUs since the requirements for different modes of operation can be very different. Moreover, it is nearly impossible to adapt a given data structure such that it possesses similar good performance on CPUs and GPUs.

1. **Analysis of IMD properties**
   IMD does not apply external libraries. This is an advantage if compared for example to pasimodo which relies heavily on external libraries (www.itm.uni-stuttgart.de/research/pasimodo/pasimodo_de.php). It turned out that it is nearly impossible to port this program to GPUs since the workload is spread quite homogeneously over all subroutines.
   IMD has been analyzed and ported partially by a group in Poland under the leadership of Rudniki [24]. IMD uses tabulated potentials and none of the special

functions which are available on GPUs. Thus it cannot take advantage of the special GPU features. IMD in general has a rather low arithmetic intensity of computations. There are only 173 instructions per byte fetched. Thus the performance limits are the low utilizability of the graphics card, redundant computations and the small number of registers available. There is, however, no memory bandwidth problem as it is often the case.

The main computation work load in IMD is the force computation which needs 80–98 % of the run time. Thus it is only one single kernel which has to be ported. After optimization, the force computation kernel requires only slightly more than 50 % of the run time.

2. **Adaption of IMD**
   A thread on the GPU is the computation task of a single atom. A thread block is a single cell of interacting atoms. The data to be sent are the coordinates, atom types, embedding energy and derivative in the case of embedded atom potentials. The received data are forces, energies, virial, and the embedding energy and its derivative.
3. **The algorithm on the GPU**
   First load the cell number, position, atom count and number and the position of neighbors. Then each thread block loads the atom data to shared memory. The threads compute interactions in the same cell. Next prefetch the data of the neighbor cells, preprocess the non-interacting neighbors. Now the threads compute the interaction with the neighbors. At last return the results to global memory.
4. **Optimization IMD**
   The data structure has to be changed from "first atom, next atom" to "first coordinate", "second coordinate" and so on to take advantage of the memory structure on the GPU. Further optimizations are given by sorting for spatial locality, applying actio equal reactio, computing the interaction matrix first and summing it up. The atoms are renumbered such that subsequent atoms are close in space. For the computation with neighboring cells the atoms are renumbered dynamically. The workflow is changed such that CPU and GPU can work in parallel.
5. **Speedups**
   The ported IMD code has been tested on several configurations, notably on 24 nodes with 2 AMD 6134 (8 Core) and 2 Nvidia 480 GTX cards, with single node Intel E 5620 (8 Core) and 2 Tesla C2050, and on single Laki nodes Intel E 5520 (8 Core) and 1 Tesla S1070 card. The raw speedup of about 100x (as predicted by Nvidia!) is reduced to 7x by geometric constraints. The speedup with respect to one CPU is 20 to 40x, with respect to the 8 Cores available it is 2.5 to 4x. Tesla and GTX are similar, and there is no notable difference between consumer and professional cars which means that CRC control does not play a role.

   The speedup is comparable to lammps which is a similar general purpose MD code. It is certainly inferior to HALMD or HOOMD for the reasons noted above.

Up to now only pair interactions and the rather similar embedded atom interactions have been ported. This means that simulations of metals are possible. These interactions do not require Verlet tables which are a major problem for the porting to GPUs, since they are realized as pointers in `IMD` and thus cannot be transferred to GPUs in a simple manner. Porting further parts of `IMD` is work in progress.

8 A Comment on World Records Molecular Dynamics Simulations

Recently the world record for atomistic MD simulations was raised to 4.125×10^{12} particles by Vrabec et al. [25]. The previous record stood at 1×10^{12} [12] for about 7 years. If a rather big time step of 1 fs is assumed for the simulation of such a sample with an edge length of about 5 μm, furthermore a typical velocity of sound of 500 m/s, then the simulation has to last of the order of 1 million time-steps to allow the sound and thus the information of the local state of the sample to cross the simulation box once! The simulations, however, were run for 50 time-steps only. In contrast, the simulations of Abraham et al. [1] in 2002 of a billion atoms were run for 200,000 time steps, thus allowing sound to cross the sample about 3–4 times during simulation.

If we look at the size of the samples we find that the only purpose of these records is to fill up the memory of the supercomputers. Continuing the previous trends we should have reached about 1×10^{14} particles already. So the memory has not grown as expected. Looking at the velocity of sound we would need at least several hundred million of time steps which means that we have to run the simulations on the **full** machine for up to a year (the simulations of Abraham et al. lasted 4 days, thus size × time-steps was the same as Vrabec et al.'s 11 years later).

We conclude that MD simulations cannot use the full power of the biggest supercomputers effectively, since the number of time-steps has to increase in accordance with the number of processing units (which leads to a larger main memory). More time-steps can only be achieved by faster processing units with higher clock rates. GPUs would certainly help to some extent, but they would not overturn the basic conclusions. Longer time-steps are not applicable, since the size of the time step is given by the basic vibration frequency of the atoms and thus by physics.

9 Summary

We have presented the molecular dynamics simulations in general as they are applied in condensed matter physics. We have introduced `potfit` to obtain accurate interactions from ab-initio calculations. A number of recent applications of IMD to multi-scale simulations have been described. We have further reported the first attempts to port IMD to GPUs.

Although there will be changes in the maintenance, porting and further development of IMD in future we are confident that it is a valuable MD simulation package and will continue to be in future.

References

1. Abraham, F.F., Walkup, R., gao, H., Duchaineau, M., Diaz De La Rubia, T., Seager, M., Simulating materials failure by using up to one billion atoms and the world's fastest computer: Brittle fracture. Proc. Nat. Acad. Sci **99** 5777–5782 (2002).
   Abraham, F.F., Walkup, R., gao, H., Duchaineau, M., Diaz De La Rubia, T., Seager, M., Simulating materials failure by using up to one billion atoms and the world's fastest computer: Work hardening. Proc. Nat. Acad. Sci **99** 5783–5787 (2002).
2. Allen, M.P., Tildesley, D.J. Computer simulations of liquids, Oxford University Press 1987.
3. Beck, P., Brommer, P., Roth, J., Trebin, H.-R., *Ab initio* based polarizable force field generation and application to liquid silica and magnesia. J. Chem. Phys. **135** 234512 (2011).
4. Beck, P., Brommer, P., Roth, J., Trebin, H.-R., Influence of polarizability on metal oxide propeties studied by molecular dynamics simulations. J. Cond. Matt. **24** 485401 (2012).
5. Brommer, P., Gähler, F., Potfit: effective potentials from ab-initio data. Modelling Simul. Mater. Sci. Eng. **15** 295–304 (2007).
6. Brommer, P., Gähler, F., Mihalcovič, M., Ordering and correlation of cluster orientations in $CaCd_6$, Phil. Mag. **87** 2671–2677 (2007).
7. Brommer, P., Beck, P., Chatzopoulos, A., Gähler, F., Roth, J., Trebin, H.-R., Direct Wolf summation of a polarizable force field for silica. J. Chem. Phys. **132** 194109 (2010).
8. Buehler, M.J., Dodson, J., van Duin A.C.T., Meulbroek, P., Goddard, W. A., The Computational Materials Design Facility (CMDF): A powerful framework for multiparadigm multi-scale simulations, Mat. Res. Soc. Proceedings (Combinatorial Methods and Informatics in Materials Science), **894**, LL3.8 (2006).
9. Elsener A. Politano, O., Derlet, P.M., Van Swygenhoven, H., Mod. Sim. Mat. Sci. Eng. **16** 025006 (2008).
10. Hocker, S., Beck, P., Schmauder, S., Roth, J., Trebin, H.-R., Simulation of crack propagation in alumina with *ab initio* based polarizable force field. J. Chem. Phys. **136** 084707 (2012).
11. Euchner, H., Mihalcovič, M., Gähler, F., Johnson, M.R., Schober, H., Rols, S., Suard, E., Bosak, A., Ohhashi, S., Tsai, A.-P., Lidin, S., Pay Gomez, C., Custers, J., Paschen, S., de Boissieu, M., Anomalous vibrational dynamics in the Mg_2Zn_{11} phase, Phys. Rev. B **83** 144202 (2011).
12. Germann, T.C., Kadau, K., Trillion-atom molecular dynamics becomes reality, Int. J. Mod. Phys. **C** 1315–1319 (2007).
13. Roth, J., Shock waves in complex binary solids: Cubic Laves crystals, quasicrystals, and amorphous solids, Phys. Rev. B **71**, 064102 (2005).
14. Roth, J., Shock waves in materials with Dzugutov-potential interactions, Phys. Rev. B **72**, 014125 (2005).
15. Roth, J., ω-phase and solitary waves induced by shock compression of bcc crystals, Phys. Rev. B **72**, 014126 (2005).
16. Roth, J., Gähler, F., Trebin, H.-R., A molecular dynamics run with 5.180.116.000 particles. Int. J. Mod. Phys. C **11**, 317–322 (2000).
17. Roth, J., Karlin, J., Sartison, M., Krauš, A., Trebin, H.-R., Molecular dynamics simulations of laser ablation in metals: parameter dependence, extended models and double pulses, in High Performance Computing in Science and Engineering '12, eds. W.E.Nagel, D.B. Kröner, M.M. Resch, Springer Heidelberg, 2013, in print.

18. Roth, J., Trichet, C., Trebin, H.-R., Sonntag, S., Laser ablation of metals, in High Performance Computing in Science and Engineering '10, eds. W.E.Nagel, D.B. Kröner, M.M. Resch, Springer Heidelberg, 2011, pp. 159–168.
19. Sonntag, S., Roth, J., Gähler, F., Trebin, H.-R., Femtosecond Laser Ablation of Aluminum, Appl. Surf. Sci. **255** 9742 (2009).
20. Sonntag, S., Trichet Paredes, C., Roth, J., Trebin, H.-R., Molecular Dynamics Simulations of Cluster Distribution from Femtosecond Laser Ablation in Aluminum, Appl. Phys. A **101** 559–565 (2011).
21. Streitz, F.H., Mintmire, J.W., Electrostatic potentials for metal-oxide surfaces and interfaces. Phys. Rev. B **50** 11996–12003 (1994).
22. Stadler, J., Mikulla, R., Trebin, H.-R., IMD: A software package for molecular dynamics studies on parallel computers. Int. J. Mod. Phys. C **8**, 1131–1140 (1997).
23. Tangney, P., Scandalo, S., An *ab initio* parametrized interatomic force field for silica. J. Chem. Phys. **117** 8898–8904 (2002).
24. Tredak, P., Lewinski, B., Ligowski, L., Wejranowski, T.,Rudniki, W.R., Large scale molecular dynamics simulations of materials on GPU clusters, Preprint
25. Vrabec, J., private communication
26. Wolf, D., Keblinski, P.,Philipot, S.R., Eggebrecht, J., Exact method for the simulation of Coulombic sysztems by spherically truncated, pairwise r^{-1} summation. J. Chem. Phys. **110** 8254–8282 (1999).

Evaluation of FastFlow Technology for Real-World Application

Kamran Idrees, Mathias Nachtmann, and Colin W. Glass

Abstract It is challenging to parallelize a real application without the knowledge of low level modes of parallelism available in the application, such as data parallelism and task parallelism. Until now, the parallel programming models focus on these low level modes of a program to exploit parallelism. FastFlow provides an alternative high level pattern based mechanism to parallelize an application. It provides pattern specific constructs to parallelize an application, in order to achieve good parallel performance and ease of programming. FastFlow has been evaluated for basic kernels. In this paper we evaluate it based on a real-world application from Molecular Dynamics.

1 Introduction

FastFlow is a C + + library used to parallelize applications efficiently by exploiting the patterns inherent to applications. It currently targets cache-coherent shared memory architectures. FastFlow provides high level constructs to explicitly define the patterns of the applications. FastFlow has been evaluated for basic kernels. In this paper we evaluate FastFlow based on a Molecular Dynamics (MD) application and compare it to an OpenMP [5] implementation.

MD simulates the interactions between molecules [4]. In principle, once the initial positions and velocities of molecules are provided, the following time progression of the molecules is deterministic. After the system is initialized, calculation of forces is done on all molecules in the system. Then, the Newton's equations

K. Idrees (✉) · M. Nachtmann · C.W. Glass
High Performance Computing Center Stuttgart (HLRS), Stuttgart, Germany
e-mail: idrees@hlrs.de; nachtmann@hlrs.de; glass@hlrs.de

M.M. Resch et al. (eds.), *Sustained Simulation Performance 2013*,
DOI 10.1007/978-3-319-01439-5_6,

of motion are integrated to advance the positions and velocities of molecules. The simulation is advanced until the computation of the time evolution of the system is completed for a specified length of time.

2 Algorithms

We used our in house Molecular Dynamics code CMD, developed for basic research into High-performance computing. CMD features multiple MD data structures, algorithms and parallelization strategies and thus allows for quantitative comparisons between them. There are two widely used data structures implemented – with corresponding algorithms – for the computation of interactions between molecules in the system. Both have been ported to FastFlow and compared to OpenMP.

2.1 BasicN2 Algorithm

In the `BasicN2` algorithm, the distance between every pair of molecules in the domain is computed. For every pair with a distance lower than the cut-off radius, the interaction is computed. It is a compute intensive and highly data parallel problem. This algorithm has the compute complexity of $O(N^2)$.

2.2 MoleculeBlocks Algorithm

In the `MoleculeBlocks` algorithm, the domain is spatially decomposed into cells (of the size cut-off radius) and then the molecules are distributed among these cells. In this algorithm, the distances between the molecules are computed only for intra-cell and neighboring cells. The `MoleculeBlocks` algorithm therefore has the compute complexity of O(N). Furthermore, Newton's 3rd law of motion is used to reduce the compute effort by half. Figure 1 shows an example of the `MoleculeBlocks` algorithm for a 2D domain space. Here the interactions of molecules of a centered cell are computed with only 4 of its neighbor cells (yellow colored) instead of all 8 neighbor cells. When the interaction between a pair of molecules is computed, the resulting force is written to both molecules. Thus, the centered cell (see Fig. 1) modifies the forces of its molecules and molecules of its right and lower neighbor cells (yellow colored).

Although the use of Newton's 3rd law in this algorithm lessens the computational effort, it raises the requirements regarding synchronization. This is because the calculated force between a pair of molecules has to be written to the data structure of both molecules.

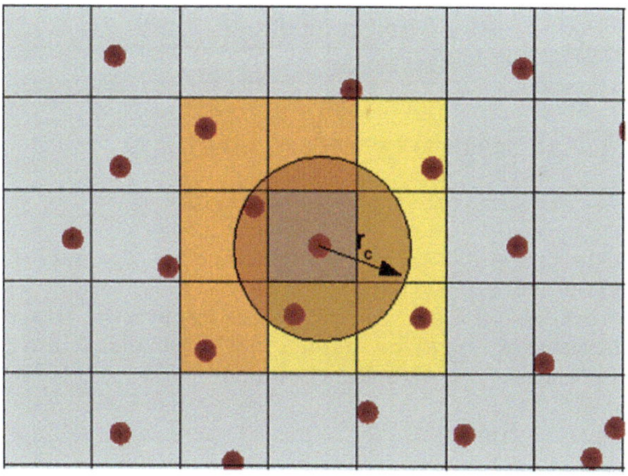

Fig. 1 Calculation of interaction between molecules using the `MoleculeBlocks` algorithm

3 Porting CMD to FastFlow

FastFlow provides powerful constructs to define patterns in an application. The two most basic constructs are pipeline and farm [1–3], whereas both of these constructs can be nested with each other. Other constructs include farm and pipeline with feedback, farm as software accelerator and map.

In a pipeline construct, multiple stages can be added as per requirement. The output of the first stage is fed to the next stage in the pipeline, which consumes its input and produce output to be used as input by the next stage and so on. Defining a pipeline allows for multiple stages to be executed in parallel on different data.

The farm pattern can be thought of as a two or three stage pipeline pattern. The first stage is an emitter which is used to offload the tasks to the second stage. The second stage is a set of worker threads which process the tasks assigned to them. The third stage is called collector and is optional. Either the second stage can directly store the result in the memory or it can feed its result to the collector, which is responsible for the reduction (or some similar operation) of the results received.

For the BasicN2 algorithm, we utilized the farm pattern in an accelerator mode. Similarly for the MoleculeBlocks algorithm, the farm pattern is used in accelerator mode and synchronization among threads is provided using locking constructs available in FastFlow.

3.1 *Parallelization of BasicN2 Algorithm*

In the BasicN2 algorithm, most of the simulation time is spent in the force calculation routine (e.g. 99.91 % for 68,000 molecules). Therefore, only the force calculation routine is parallelized. A farm is created with the given number of

Algorithm 1 Pseudo code for *Farm* creation

```
int main(int argc, char *argv[]) {

    /* Parsing of input arguments - identifies number of
        worker threads */
    ff_farm<> accelerator(true); /* Declare FastFlow Farm
        in Accelerator Mode */
    std::vector<ff_node *> workersVector; /* Declare vector
        of Farm worker threads */
    /* Add workers to Farm in the grid generator routine */
    grid_generator(&target_ensemble, SC, &domain, psp,
        &accelerator, &workersVector);
    ...
    start_timer = timer();
    Main_Simulation_Loop {
        Pre-Force Integration();
        Force-Calculation(&accelerator); /* Parallelized
            using farm in accelerator mode */
        Post-Force Integration();
        ...
    }
    end_timer = timer();
return 0;
```

worker threads before the simulation is started. The number of threads is controlled using the command line argument, '–num-threads'. The farm is used in the accelerator abstraction to parallelize only the part of the simulation which computes forces on molecules.

Once the simulation is started, inside the basicN2_calc_forces routine, the farm is launched and the molecules are divided among its worker threads (by an emitter thread) to compute the forces acting on them. Each thread computes the forces and accumulates the potential energy of its molecules. The main thread is frozen to wait for the worker threads to finish their tasks. Once all threads have finished, a reduction operation is performed by the main thread to calculate the global sum of potential energy of all the molecules in the system. To avoid the run-time overhead of creating the farm again and again, it is not destroyed, but reused in every iteration. This is done using the wait_freezing() method of the farm. Algorithm 1 shows the pseudo code of creating the farm in accelerator mode and Algorithm 2 shows the pseudo code of parallelizing the force calculation routine using the farm construct.

3.2 Parallelization of MoleculeBlocks Algorithm

As shown in Table 1, most of the simulation time is spread among the multiple routines for the MoleculeBlocks algorithm.

Algorithm 2 Use of *Farm* pattern in *BasicN2*

```
void basicN2_calc_forces(void *container, real *U_pot,
        ff_farm<> *accelerator) {
    basicN2 *mc = (basicN2 *) container;
    real U_pot_tmp = 0.;

    if(accelerator->run_then_freeze() < 0) {
        error("running accelerator\n");
    }

    for(int i = 0; i < num_threads; i++)
        accelerator->offload(mc); /* Broadcast pointer to
            molecule container to all threads */

    accelerator->offload((void *) FF_EOS);
    accelerator->wait_freezing(); /* Freeze farm */

    /*Reduction Operation for global sum of Potential Energy*/
    for(i = 0; i < config.num_threads; i++)
        U_pot_tmp += U_pot_worker[i];

    *U_pot += U_pot_tmp;
}
```

Table 1 Profile of CMD *Molecule Blocks* use case

Routine	Time spent on routine (%)
Force calculation	41.74
Kinetic energy calculation	15.36
Pre-force integration	10.43
Post-force integration	4.22
...	...

Therefore to achieve good speedups, all expensive routines of the simulation are parallelized. Similar to the parallelization of the BasicN2 algorithm, a FastFlow farm is created before the actual simulation starts. Then, the main thread launches the farm inside all of the expensive routines of the simulation. As the farm is launched, the emitter thread distributes the cells among the worker threads to perform the parallelization of a task specific to the routine to be parallelized. After starting the farm, the main thread waits for the worker threads to finish their tasks, using the wait_freezing() routine. The different simulation routines must be executed serially, however, each routine can be parallelized. Here, the same farm is reused by all parallel routines. This is done by passing a task-pointer to the worker (or service) routine of the farm. The pointer points to a structure, which has a member identifying which routine is currently to be executed in parallel by the farm (as shown in Algorithm 3). Inside the worker (or service) routine, switch cases

Algorithm 3 Service routine of worker threads in *MoleculeBlocks* algorithm

```
class Worker: public ff_node {
public:
    void * svc(void * ff_task) {
        ff_task_t * task = (ff_task_t * )ff_task;
        mb_t *mc = (mb_t *) task->container;
        int threadID = ff_node::get_my_id();

        if(CALC_FORCES == task->routine) {
            /* This routine is parallelized here */
            /* Distribution of molecule cells is
               done in a manner that each thread
               has contiguous cells in all
               dimensions — it reduces the
               probability of a thread to wait
               for acquiring lock on neighbor
               cell for force calculation*/
        }

        else if (INTEGRATE_PREF == task->routine) {
            /* This routine is parallelized here */
        }

        else if (INTEGRATE_POSTF == task->routine) {
            /* This routine is parallelized here */
        }

        else if (CALC_E_KIN == task->routine) {
            /* This routine is parallelized here */
        }

        ...

        return GO_ON;
    }
};
```

are used to determine which routine is currently to be executed. After all threads have finished computing their current task, the main thread performs a reduction operation to calculate global parameters of the complete system.

Synchronization in the force-calculation routine (in order to avoid race conditions) is performed by pthread locks. The pthread_mutex_trylock routine is used to acquire the locks on the two cells between which the molecular interactions are to be computed. The use of this non-blocking lock acquiring routine avoids deadlocks among threads. If a thread acquires the lock on a cell and it does not get the lock on its neighbor cell, it releases the acquired lock.

4 Evaluation

The CMD BasicN2 algorithm is evaluated for the problem size of 68,000 molecules, whereas the CMD MoleculeBlocks algorithm is evaluated for the problem size of 1,000,000 molecules. Both codes are compiled with GCC 4.7.2. Furthermore, for the FastFlow implementation, threads are pinned with the following mapping: main thread to core 0, emitter thread to core 1, worker threads to cores 1, 2, 3, 4, 5, 6, 7, 8, 9, 10, 11, and 0, respectively (hyperthreading).

4.1 Evaluation Metrics

Execution Time

This is the total time spent in the simulation loop of CMD, which incorporates 7–8 routines. The execution time is averaged over 10 independent simulation runs. The benchmark results show the average values and standard deviations of the execution time.

Speedup

This is the speedup of the simulation loop when executed in parallel. It is the ratio between the average serial execution time and the average parallel execution time of the simulation loop.

4.2 Results

Execution Time

BasicN2 Algorithm

The benchmark for the BasicN2 execution time is shown in Fig. 2. It compares the execution time of CMD BasicN2 parallelized with FastFlow and OpenMP. We can see that the OpenMP version is slightly faster than the FastFlow version for small numbers of threads (as shown in Table 2). But as the number of threads is increased, we achieve almost the same execution time.

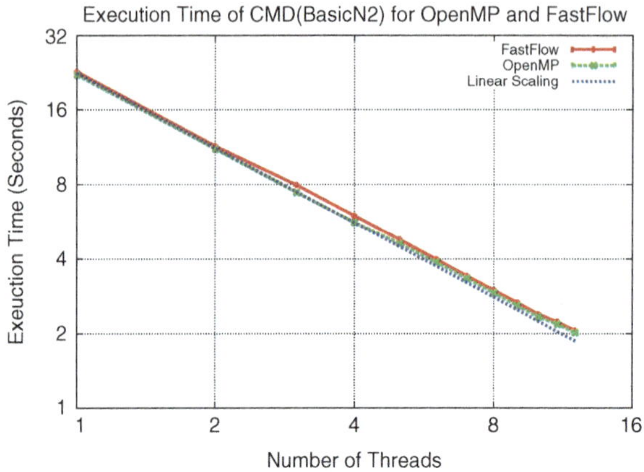

Fig. 2 Execution time of `BasicN2` algorithm with FastFlow and OpenMP

Table 2 Averaged measured execution times and standard deviation for the BasicN2 and MoleculeBlocks algorithms implemented with FastFlow (FF) and OpenMP, respectively

	BasicN2		MoleculeBlocks	
	FF	OpenMP	FF	OpenMP
cores	time (s)	time (s)	time (s)	time (s)
1	22.819 ± 0.011	22.126 ± 0.008	2.335 ± 0.002	2.364 ± 0.003
2	11.435 ± 0.026	11.132 ± 0.025	1.442 ± 0.004	1.659 ± 0.031
3	7.978 ± 0.012	7.443 ± 0.013	1.090 ± 0.002	1.252 ± 0.113
4	5.985 ± 0.021	5.631 ± 0.054	0.926 ± 0.001	1.088 ± 0.027
5	4.802 ± 0.014	4.621 ± 0.003	0.819 ± 0.001	0.903 ± 0.008
6	3.993 ± 0.006	3.885 ± 0.005	0.923 ± 0.005	0.849 ± 0.001
7	3.417 ± 0.006	3.337 ± 0.007	0.792 ± 0.001	0.753 ± 0.001
8	2.997 ± 0.008	2.921 ± 0.006	0.759 ± 0.002	0.745 ± 0.002
9	2.667 ± 0.010	2.598 ± 0.007	0.733 ± 0.001	0.697 ± 0.017
10	2.399 ± 0.008	2.340 ± 0.003	0.681 ± 0.003	0.691 ± 0.013
11	2.234 ± 0.001	2.190 ± 0.001	0.671 ± 0.001	0.648 ± 0.001
12	2.064 ± 0.001	2.022 ± 0.001	0.720 ± 0.004	0.673 ± 0.001

MoleculeBlocks Algorithm

The benchmark for the MoleculeBlocks execution time is shown in Fig. 3. It compares the execution time of CMD MoleculeBlocks parallelized with FastFlow and OpenMP. We can see that the FastFlow version is faster than the OpenMP version for up to 5 worker threads (single NUMA Node on RGU Cluster has 6 cores). The FastFlow implementation uses a farm in accelerator mode. It has two additional threads (main and emitter thread) compared to OpenMP. FastFlow utilizes one core for the main thread, whereas the first worker thread is mapped to the same core as the emitter thread (hyperthreading). So for 5 worker threads the FastFlow implementation utilizes one complete NUMA node of the utilized RGU

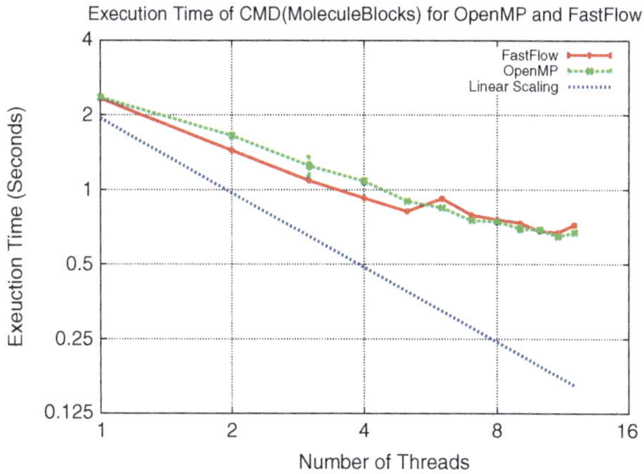

Fig. 3 Execution time of `MoleculeBlocks` algorithm with FastFlow and OpenMP

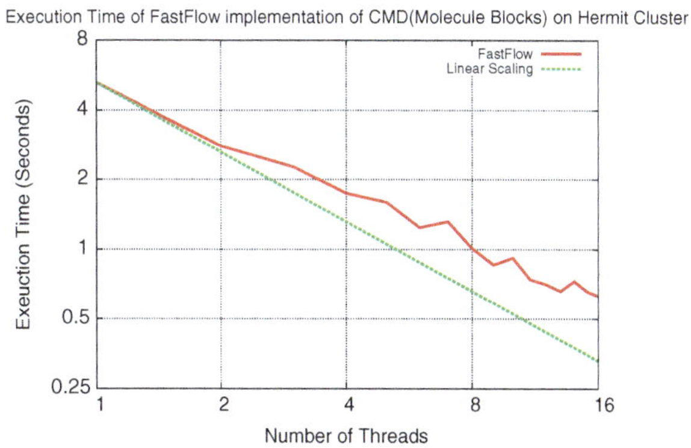

Fig. 4 Execution time of `MoleculeBlocks` algorithm with FastFlow with cut-off radius = 3

cluster (6 cores per NUMA node, 12 per node). As the number of worker threads is increased beyond 5, the OpenMP version performs slightly better than the FastFlow version.

By increasing the cut-off radius, we can control the computation to communication ratio and hence can explore the behavior of the parallel implementations in more detail. Figure 4 shows a benchmark for the execution time of MoleculeBlocks algorithm with the cut-off radius of 3.

Speedup

BasicN2 Algorithm

The benchmark for the speedup of BasicN2 is shown in Fig. 5. We observe almost the same speedup with FastFlow as we do with OpenMP.

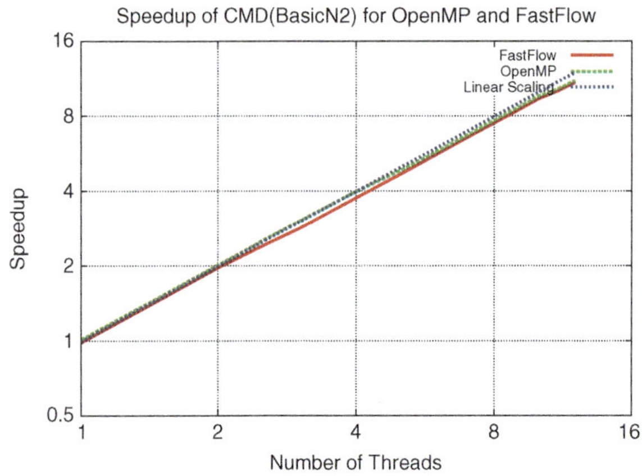

Fig. 5 Speedup of `BasicN2` algorithm with FastFlow and OpenMP

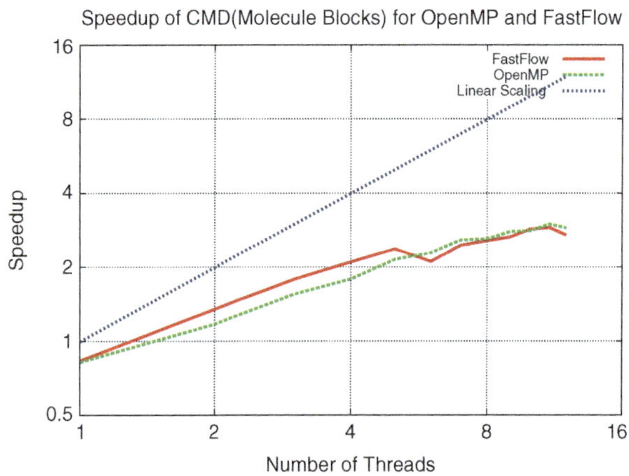

Fig. 6 Speedup of `MoleculeBlocks` algorithm with FastFlow and OpenMP

MoleculeBlocks Algorithm

The benchmark for the speedup of MoleculeBlocks is shown in Fig. 6. We observe that the FastFlow version is significantly faster on a Single NUMA node (as shown in Table 2). However, beyond a single NUMA node, the OpenMP version is slightly faster. Figure 7 shows the benchmark for the speedup of the MoleculeBlocks algorithm with the cut-off radius of 3.

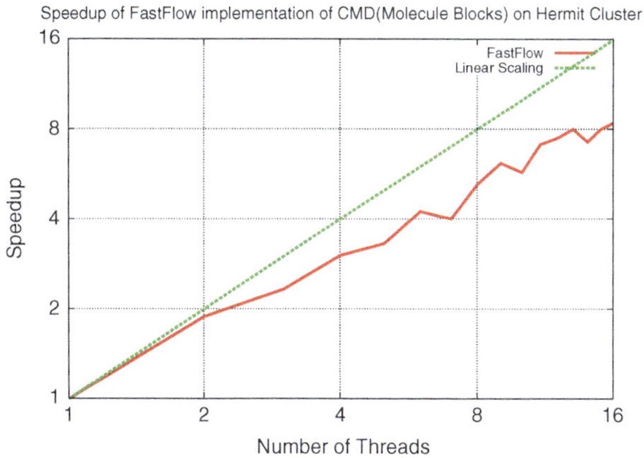

Fig. 7 Speedup of FastFlow implementation of `MoleculeBlocks` algorithm with cut-off radius = 3

5 Conclusions

With both algorithms, BasicN2 and Molecule Blocks, we have observed very similar speedups when comparing FastFlow and OpenMP implementations. However, there is still a need to efficiently map the threads to the cores when using FastFlow on Shared Memory Multiprocessor (SMP) systems. In an SMP system with more than a single Non-Uniform Memory Access (NUMA) node, like the RGU cluster, the placement of a thread and the data it accesses can have a severe impact on the performance of an application. This is because in SMP with NUMA nodes, each NUMA node has its separate memory interface. Therefore, if a thread is pinned to one NUMA node while the data it access is located on another, it has high memory access latency, compared to data residing on the same NUMA node. This effect is apparent in Figs. 3 and 6. As we understand, there are plans to implement an automatic efficient data mapping in FastFlow. The MD use case could profit significantly from such a mapping. Further developments of increasingly high level patterns are also planned by the developers. The presented MD MoleculeBlocks use case would benefit especially from a high level stencil pattern. The implementation of this pattern should include implicit synchronization strategies, when an elemental function accesses neighbor elements. In this way, more efficient synchronization schemes could be achieved.

Acknowledgements This work was funded by the EU Project Paraphrase.

References

1. Marco Aldinucci, Marco Danelutto, Peter Kilpatrick, Massimiliano Meneghin, and Massimo Torquati. Accelerating sequential programs using fastflow and self-offloading. *arXiv preprint arXiv:1002.4668*, 2010.
2. Marco Aldinucci, Marco Danelutto, Peter Kilpatrick, and Massimo Torquati. Fastflow: high-level and efficient streaming on multi-core. *Programming Multi-core and Many-core Computing Systems, Parallel and Distributed Computing*, 2012.
3. Marco Aldinucci, Marco Danelutto, and Massimo Torquati. Fastflow tutorial. *arXiv preprint arXiv:1204.5402*, 2012.
4. Michael P Allen. Introduction to molecular dynamics simulation. *Computational Soft Matter: From Synthetic Polymers to Proteins*, 23:1–28, 2004.
5. Leonardo Dagum and Ramesh Menon. Openmp: an industry standard api for shared-memory programming. *Computational Science & Engineering, IEEE*, 5(1):46–55, 1998.

Storage and Indexing of Fine Grain, Large Scale Data Sets

Ralf Schneider

Abstract In this work a storage scheme for fine grain data sets, which are large in absolute size as well as in file number, is presented. It is meant to deal in parallel with data arising from simulation tasks utilizing high performance computing systems for the independent parallel execution of several thousand serial tasks, resulting in large data sets of terabyte size spread over hundreds of thousands or even millions of files.

First of all a short introduction is given to the problem of the multiple parallel execution of serial programs on high performance computing systems. This leads to the description of the idea how to realize a data storage and indexing concept to handle fine grain, large scale data sets. Afterwards the implementation in form of a FORTRAN 2003 library is presented along with the use cases evaluated in the results section.

1 Introduction

When speaking of large data sets in high performance computing two major categories are separated in this work:

1. **Coarse grain** data sets arising from large integrated parallel simulations on large grids generating several terabytes (TB) of result data aggregated in only a few files.
2. **Fine grain** data sets arising from tasks where small, independent, mostly serial simulations are carried out several thousand times. Even though the amount of data generated by these tasks is the same as in the first category there is the

R. Schneider (✉)
High Performance Computing Center Stuttgart, Nobelstraße 19, 70569 Stuttgart, Germany
e-mail: schneider@hlrs.de

M.M. Resch et al. (eds.), *Sustained Simulation Performance 2013*,
DOI 10.1007/978-3-319-01439-5_7,

significant difference that the data are spread over hundred thousands of files each heaving a size of only several megabytes (MB) up to a few Gigabytes (GB).

Typical simulations which generate data of the first category are for example CFD-Simulations of transient processes or multiphysics simulations like Fluid-Structure-Interaction. The codes which are utilized to carry out these simulations are typically highly parallel codes which are optimized to run in a nearly ideal manner on high performance computers.

The tasks which generate data sets of the second category are namely parametric studies, optimization tasks or, as in the case of the presented work, procedures from which functional dependencies between different physical scales are derived [1, 2].

The first basic characteristic of these tasks is in most cases, that not only a single simulation code takes part in the analysis of the given problem but multiple programs, which are not closely integrated, are utilized in a kind of process chain to derive the desired simulation result. The second characteristic is that the process chain is executed ten thousands up to millions of times with different input data where each execution produces its own set of output data organized in files of small size.

If a fine grain data set has to be post processed after its creation which means reorganized, scanned for certain data or simply being moved for the sake of feature extraction, process evaluation or data backup certain problems arise resulting from the given data set layout in combination with the performance parameters of modern high performance storage systems like Lustre.[1] The aim of this paper is to discuss these problems and describe a possible solution for them with help of a parallel storage and indexing scheme for fine grain data sets resulting from large scale simulations utilizing high performance computers.

1.1 Description of Use Case

Since the ideas presented in this work grew during the development of a simulation task for the calculation of continuum material data for micro-structured materials and because this task is used in this work as a prototype for the generation of fine grain data sets its basic implementation principles are described in this section.

As shown in Fig. 1 the basis for a single instance of the process chain is a high resolution volume data set from which in the first step a FE-Geometry of its internal micro-structure is extracted along with definitions of different load case scenarios. In the second step the mechanical response of the generated geometry due to the different load cases is evaluated by means of the FE-Package FMPS.[2] From the output data produced by the load case simulations and collected during the third

[1] For detailed information see www.lustre.org

[2] FMPS is provided by Dr. Hans Wüstenberg and LASSO Ingenieurgesellschaft mbH

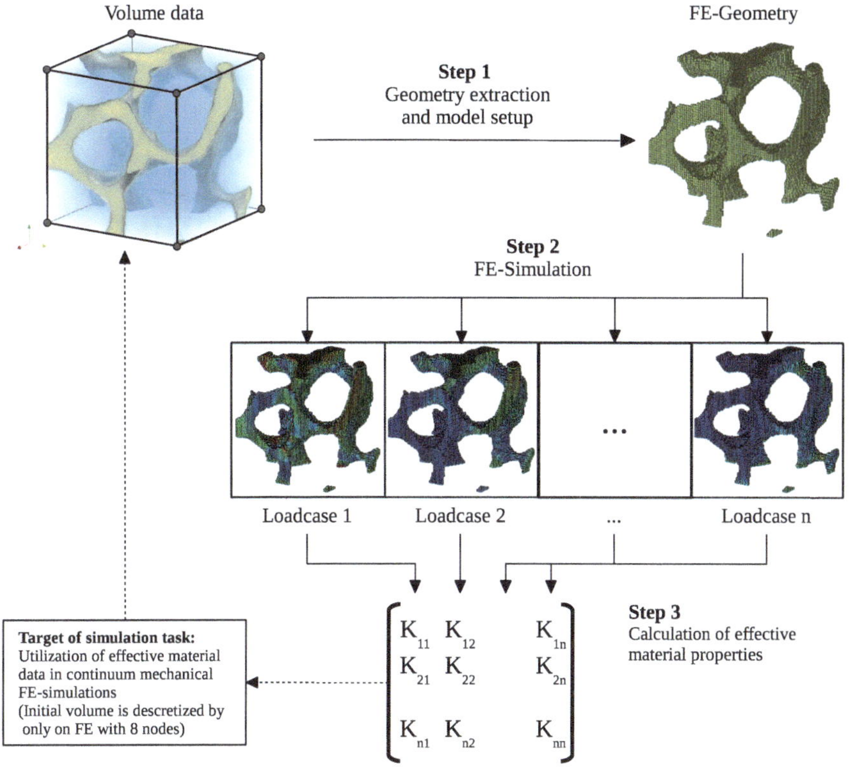

Fig. 1 Process chain to extract continuum material data of micro-structured materials by direct numerical simulation

step of the process chain, the effective material properties of the extracted micro structure are calculated.

The effective material data derived by this process describe the behavior of the micro-structure enclosed by the initial volume data in a way that the complete volume can now be discretized on the continuum mechanical scale by only one homogeneous, brick shaped finite element with 8 nodes. In other words, the generated material data can be utilized in continuum mechanical FE-simulations so enabling a very detailed local structural analysis in very short time and with very less computational effort.

The question why the described process chain has to be executed ten or hundred thousands of times is answered in Fig. 2 where a data set of a human femoral head is shown along with several samples of micro structure extracted from it. Since human bone tissue is a naturally growing material it can be easily imagined that no matter how the shown data are decomposed into base data sets for the process chain described above no two data sets can be found which contain the same micro-structure. This means if the complete data set of the femoral head is decomposed in

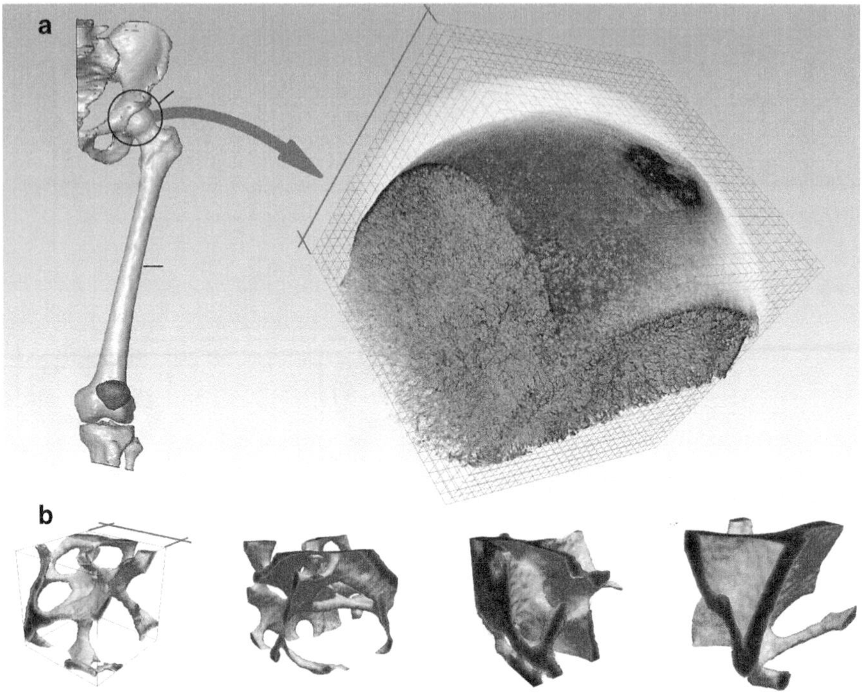

Fig. 2 Decomposition of a high resolution volume data set of a human femoral head into sub-volumes of 1.2 mm edge length

a grid of sub-volumes like shown in Fig. 2 the process chain has to be executed on each sub-volume to derive its effective material data. In this way the complete field of continuum material data for the femoral head can be calculated.

1.2 Data Amount, File Numbers and Sizes

The volume data set of a human femoral head as shown in Fig. 2 is stored in one file which requires 20 GB of storage space. During step 1 of the process chain sub-volumes of the desired size are directly loaded from this file by the geometry extraction tool. This tool then generates one model file which contains the FE-geometry description of the micro structure enclosed in the considered sub-volume along with the 24 load case description files which are necessary to evaluate the mechanical behavior of the micro-structure. Additionally a description of the sub-volume is generated which consists out of two files, describing the location of the sub-volume under consideration with respect to the decomposition of the volume data set of the complete femoral head.

Table 1 Domain count, data amount, file count and average file sizes for decompositions of the femoral head data set in sub-volumes of 0.6 and 1.2 mm edge length

Edge length	0.6 mm	1.2 mm
Domain count	169,344	21,168
Data amount [GB]	6,503.5	5,445.0
File count	4,570,470	607,650
Average file size [MB]	1.42	8.96

Subsequently the model and load case files are loaded by the FE-Package FMPS to simulate the mechanical response of the micro-structure, generating 4 files which contain the collected results of all 24 load case simulations.

The calculation of the effective material properties is afterwards carried out during step 3 loading the four result files of the preceding step, generating two result files containing the effective material data for the selected sub-volume. In summary 35 files are generated as a minimum during the execution of a single instance of the process chain including a log and a monitor file.

The domain numbers, data amount, file numbers and average file sizes resulting from a decomposition of the femoral head data set into sub-volumes of 0.6 and 1.2 mm edge length respectively are given in Table 1.

It should be noted that the total file size given in Table 1 cannot be derived by multiplication of the total domain number by 35 since not all sub-volumes contain micro structure. This is due to the fact, that the spherical shaped structure of the femoral head is contained in a cube shaped data set.

1.3 Serial Applications and Fine Grain Data Sets

To give an idea of the problems arising, when operating with serial applications on fine grain data sets, the execution times of the archiving algorithm tar[3] were measured for two representative data sets on the HLRS Cray-XE6 system. The parameters of the file sets along with the results of the measurements are given in Table 2. It can be seen that the number of processed files per second as well as the data rate in GB/s are somewhat far away from the numbers which are reported for modern Lustre file systems [3]. This is explainable by the fact, that Lustre is designed to be accessed from high performance computers with thousands of cores in parallel ideally operating on coarse grain data sets.

Tests with other single core applications, carried out during the development the described simulation process, showed the same results as tar. Thus it can be stated that the dominating factor for the execution time of single core applications

[3]http://www.gnu.org/software/tar/tar.html

Table 2 I/O performance of tar archiving a fine grain data set generated by the described simulation process

Edge length	0.6 mm	1.2 mm
Domain count	4,096	512
Data amount [GB]	265	230
Average data [GB/Domain]	0.0647	0.449
File count	167,952	20,994
Walltime	18:48:52	02:06:26
Minimal file size [MB]	16^{-6}	16^{-6}
Average file size [MB]	1.6	11.2
Maximal file size [MB]	≈ 65	≈ 500
Data rate [Files/s]	2.48	2.77
Data rate [GB/s]	0.004	0.03

operating on fine grain data sets is the number of files which can be processed per second by a single core.

Furthermore the tests showed that this number depends on the application, the data set layout, the workload other users introduce to the file system and the variation of the file size within the data set but that the dependency is in the range of less than one order of magnitude.

2 From the Reduction of Input/Output Operations Per Second (IOPS) Towards a Storage and Indexing Concept for Fine Grained, Large Scale Data Sets

When the development of the simulation process described in the previous section was started relatively little attention was payed to its I/O parameters. This changed once the number of tasks executed in parallel was raised to larger numbers. The first problem related to I/O which came up was the overloading of the Lustre Meta-Data server. This problem was induced by the number of files generated by a single instance of the material data calculation. E.g. one single load case simulation generated six files at this implementation stage which in addition were partially ASCII formatted.

This problem is also known from other centers and applications. Shipman et al. [3] reports from problems with an access pattern to a single input file which was initially done by a serial application. Once the application was parallelized the same access pattern was done by all parallel processes which also lead to an overloading of the Meta-Data servers.

2.1 Data Storage Concept

To overcome this problem the first optimization task was to reduce the number of files generated by the process to take load from the Meta-Data server. The solution for this task is based on the idea that data stored in an I/O system should rather be organized according to their data type than in logical or semantic groups. Taking again the example of the load case simulations this means that the simulation data which consist basically out of lists of integer and floating point numbers and some character data are stored in only three files. One file containing 8 Byte Integer data like external node numbers, external element numbers and element connectivity lists, a second file containing all 8 Byte Floating point data like node coordinates and element results and a third file containing all the needed ASCII data.

This storage technique has two major advantages. The first is the reduction of file create and open operations during creation and read back of the data and the second is the reduction of read operations. Since the three files are now used to store the data from all 24 load case simulations the data can be loaded back to memory by the following application with only three unformatted read operations. Because the data are now organized the same way in both, the memory and the I/O system we will call them streams in the following, so referring to either the files located on the I/O system or the data arrays located in memory.

2.2 Data Indexing Concept

To make the proposed storage concept applicable in a universal way, what is of course needed is a technique to transfer the parameters of the data chunks contained in the streams to a successive application. For that purpose an indexing concept was developed whose basic principles are:

1. No semantic information about the stored data should be included
2. Only positional and data type information should be included
3. It should be possible to group information
4. The operations extension, combination, inclusion and split should be possible

The idea behind the first two principles is to keep the index itself small and fast to handle since the number of items tracked by it is supposed to grow very fast. The parameters actually needed to describe a data item within a stream are its data type so specifying the corresponding stream, its size and its starting position within the stream.

The third and fourth principle are conditional upon the intention to enable the creation and extension of a coarse grain data set whilst preserving the fine granularity of the included data by means of the indexing scheme. For that purpose it has to be possible to organize the index in a tree like structure to be able to imitate the structure of a file system's directory tree.

In summary the concepts stated above led to an indexing scheme based on two entities one being used to hold the actual parameters of an indexed item, called a leaf and another being used to structure the indexed items called a branch. From the implementation's point of view one leaf is a set of three 64 Bit integer numbers together with an item that identifies the leaf. A branch like a directory in a file system has to be a recursive structure which can but has not to contain an arbitrary number of leaves and other branches.

2.3 Implementation

The implementation of the indexing scheme is done in FORTRAN 2003. The derived data types which represent the proposed leaf and branch entities are declared as follows:

```
Type tLeaf

  Character(Len=pd_mcl) :: desc

  Integer(Kind=1)       :: dat_ty

  Integer(Kind=pd_ik)   :: dat_no
  Integer(Kind=pd_ik)   :: lbound

End Type tLeaf

Type tBranch

  Character(Len=pd_mcl)                    :: desc

  Integer(Kind=pd_ik)                      :: no_branches
  Integer(Kind=pd_ik)                      :: no_leaves

  Type(tBranch), Dimension(:), Pointer :: branches => null()
  Type(tLeaf)  , Dimension(:), Pointer :: leaves   => null()

  Type(tStreams), Allocatable              :: streams

End Type tBranch
```

To couple the data storage concept with the indexing concept and to enable the streams to either being represented as Arrays of basic data type in memory or as files on the I/O system a third derived data type called tStreams has to be introduced. It is declared as stated below.

```
Type tStreams

  Integer(Kind=pd_ik)   , Dimension(no_streams) :: dim_st

  Character(len=pd_mcl), Dimension(no_streams) :: stream_files
  Logical               , Dimension(no_streams) :: ifopen
  Integer               , Dimension(no_streams) :: units

  Integer(kind=1), Dimension(:), pointer :: int1_st  => null()
  Integer(kind=2), Dimension(:), pointer :: int2_st  => null()
  Integer(kind=4), Dimension(:), pointer :: int4_st  => null()
  Integer(kind=8), Dimension(:), pointer :: int8_st  => null()
  Real(kind=8)   , Dimension(:), pointer :: real8_st => null()
  Character      , Dimension(:), pointer :: char_st  => null()

End type tStreams
```

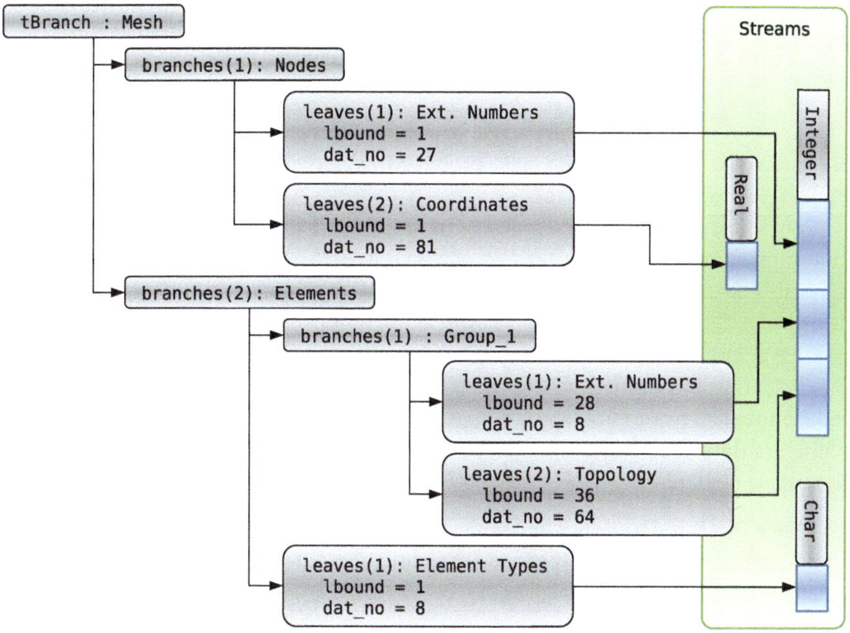

Fig. 3 Example for the data organization of a simulation mesh using the proposed storage and indexing scheme

The data type is only used as an attribute to tBranch and is not intended to be used for the declaration of stand alone variables. To give an idea to the reader how the data types can be used, the storage and indexing of a simulation result is shown in Fig. 3.

To enable the structural operations proposed in the previous section the following subroutines exist in the current implementation:

- `raise_tree`
  Initialize the root of an index structure
- `raise_branch`
  Initialize a branch
- `raise_leaves`
  Initialize a leaf
- `add_branch_to_branch`
  Add a branch to an existing branch (extend the attribute branches by one element)
- `add_leaf_to_branch`
  Add a leaf to an existing branch (extend the attribute leaves by one element)
- `include_branch_into_branch`
  Include an existing index structure to another one

3 Implementation of a Parallel Indexing and Packaging Algorithm by Means of the Proposed Concept

In this section the MPI[4] parallel implementation of an indexing and packaging algorithm for fine grain data sets, produced by the simulation process presented in Sect. 1.1, is described.

The implementation was mainly driven by the idea to avoid as much MPI communication as possible. Since the massive I/O operations of a packaging algorithm were expected to slow down the performance of the process it was aimed to reduce other performance critical operations to a minimum. To pursue this objective the developed index and storage concept is perfectly fitted. By separating the indexing and storage of the data set the algorithm can be divided into six independent phases including start up and finalization. The operations done in every phase are described below and marked with corresponding names in the algorithm flow chart depicted in Fig. 4.

- **Phase 0: Initialization** The application is initialized which means the steering parameters are loaded by process 0 and broadcasted to the other processes. In this phase a local index is initialized by every process.
- **Phase 1: indexing**
  During that phase each process traverses independently from its partners a portion of the data set. In the considered use case the work distribution between the processes is chosen to be of even portions. This can be done since load inbalance isn't considered to be much of problem due to the fact, that the number of work packages (directories) is much higher than the number of processes.
- **Phase 2: Offset communication and local index correction**
  Once the indexing is finished the so called offset communication and local index correction has to be done. This simply means that each process except the last one communicates the size of the data indexed by it to its successors. After having received all index sizes from its predecessors each process sums up the received data to a local offset. Now the local index can be corrected by traversing it and adding the local offset to the attribute `lbound` of each leaf so creating a distributed global index.
- **Phase 3: Storage**
  In this phase at first each process opens the global archive files for direct access by means of MPI-I/O. Then the work packages already assigned to the process during phase 1 are cycled again but this time all indexed data are in sequence loaded to memory and being written to the global archive files by MPI-I/O direct access.
- **Phase 4: Global index combination**
  After heaving stored all indexed data process 0 gathers the local indicees from its partners and combines them to a global index.

[4]MPI – Message-Passing Interface: http://www.mpi-forum.org/docs/mpi-3.0/mpi30-report.pdf

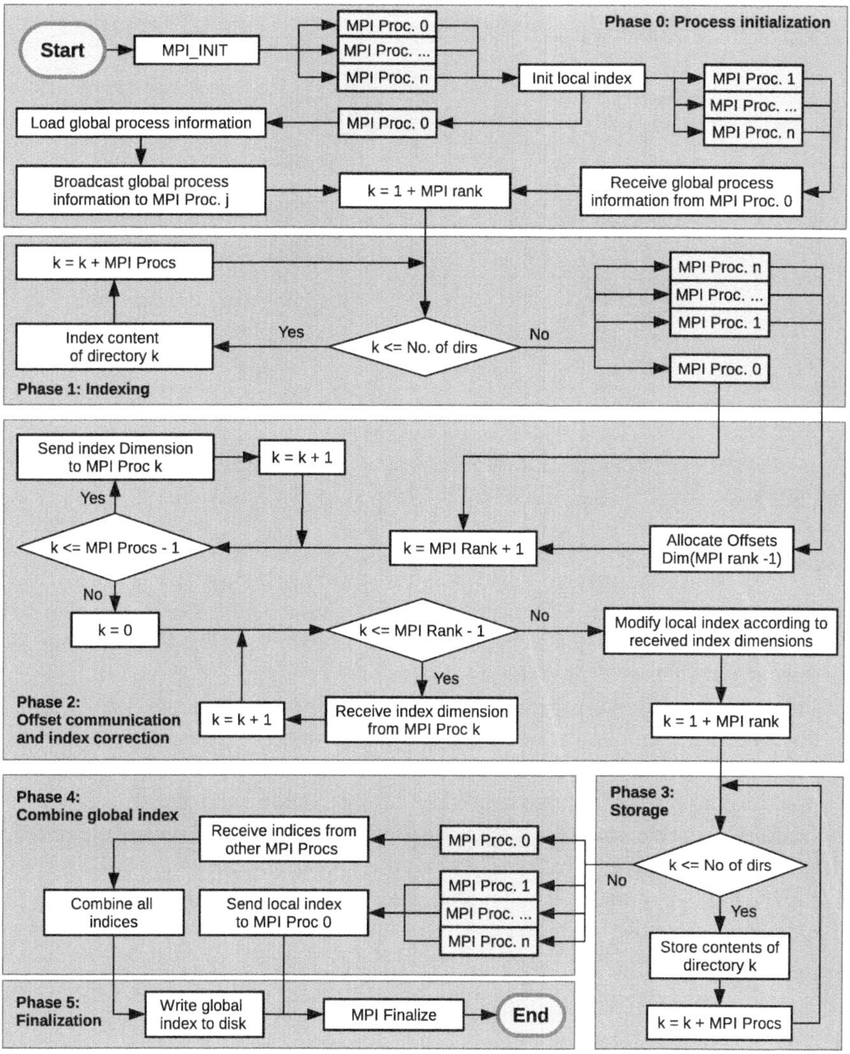

Fig. 4 Implementation of a MPI parallel packaging algorithm for fine grain data sets

- **Phase 5: Finalization**
  In this phase all processes are terminated after process 0 having stored the global index on disc.

Even though it was originally developed especially for the given data set structure it can be stated that the presented implementation led to certain principles which are of general purpose for the development of parallel packaging algorithms for fine grain data sets.

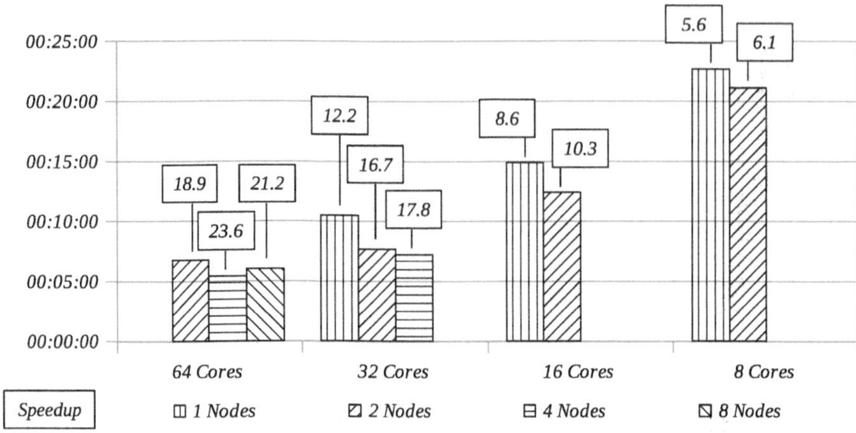

Fig. 5 Walltimes and Speedup for MPI packaging of the representative data set with 1.2 mm decomposition

1. To reduce I/O system workload it is of advantage to use a hashed directory structure. In that way each parallel process can calculate the direct path to a certain directory without having to ask a master process or do operations like directory listings affecting the I/O system.
2. Indexing and storage of the data should be split into separate phases with a concentrated communication phase in between and one at the end. In that way the necessity of a index coherence protocol can be avoided which means inter process communication isn't necessary during indexing and storage phase.
3. By fully using each process's memory for the aggregation of several elements of the fine grain data set a big amount of write operations can be avoided so taking load of the I/O system and better using the characteristics of HPC I/O systems to efficiently write big I/O buffers.

4 Application Results

In this section the application details of the presented implementation are discussed. The performance of the implementation was evaluated on Hermit the HLRS Cray-XE6 system,[5] using the representative data sets described in Sect. 1.3. In addition the packaging of data sets produced by production runs of the material data calculation process for decompositions in 0.6 and 1.2 mm cubes is analyzed. The results for the performance tests are given in Fig. 5 for the 1.2 mm decomposition and in Fig. 6 for the 0.6 mm decomposition.

---

[5] http://www.hlrs.de/systems/platforms/cray-xe6-hermit/

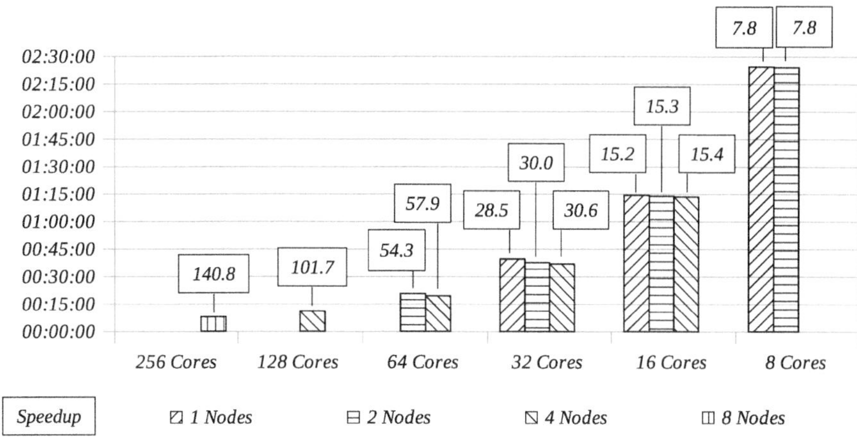

Fig. 6 Walltimes and Speedup for MPI packaging of the representative data set with 0.6 mm decomposition

In Fig. 5 the walltime for the MPI parallel packaging of the representative dataset with 1.2 mm decomposition is shown, executed with 8, 16, 32 and 64 AMD Interlagos compute cores placed on 1, 2, 4, or 8 compute nodes. The execution with the same number of cores but placed on different numbers of nodes was done to evaluate whether limitations in terms of I/O bandwidth per node exist.

As can be seen in Fig. 5 for the sake of pure speedup it can be beneficial to run the application on less than 32 cores per node which of course wouldn't be done for real applications where saving absolute walltime has the highest priority. Regarding the I/O bandwidth per node especially by the measurement taken with 64 cores it is shown that this limit isn't reached for data sets of the evaluated structure otherwise a speedup has to occur when increasing the node count from 4 to 8. An additional evidence for this statement was given by a second test carried out during the development of the application where every process wrote to its own set of archive files and the communication of the serialized local indicees was not executed. In this test the I/O bandwidth per node was approximately two times the one compared to the current implementation.

Although not finished from the analysis done so far it can already be said that an additional performance increase should be possible by optimizing the MPI communication of the serialized headers which isn't done in an optimal way by the current implementation. For that reason the evaluation shown in Fig. 6 for the 0.6 mm decomposition was done without the communication of the serialized indicees to better reveal the I/O behavior in the walltime measurements.

Looking at Fig. 6 the first feature to notice is again that no significant speedup could be gained spreading 16, 32 or 64 processes across more than the minimal number of compute nodes.

Table 3 I/O performance of MPI parallel packaging of fine grain data sets resulting from material data calculations of a human femoral head with 0.6 and 1.2 mm decomposition

Edge length	0.6 mm	1.2 mm
Domain count	169,334	21,168
Data amount [GB]	6,503.5	5,445.0
Average data [GB/Domain]	0.0384	0.257
File count	4,570,470	607,650
Walltime	02:59:42	02:01:34
No. of cores	256	512
No. of nodes	16	16
Minimal file size [MB]	16^{-6}	16^{-6}
Average file size [MB]	1.42	8.96
Maximal file size [MB]	$\approx$65	$\approx$500
Data rate [Files/s]	423.9	83.3
Data rate [GB/s]	0.603	0.747

By comparing the speedup numbers to the ones of the 1.2 mm decomposition a significant improvement can be noticed. Where this improvement actually comes from which means why smaller files can be processed more efficiently than bigger ones couldn't be finally tracked down but is currently under evaluation.

In Table 3 the performance numbers for the packaging of two data sets resulting from real applications of the material data calculation process described in Sect. 1.1 are shown. Again one data set resulted from a decomposition of a femoral head micro-CT with 0.6 mm domain edge length and the other one from a decomposition of the same micro-CT with 1.2 mm domain edge length.

It can be seen that the data set resulting from the 1.2 mm decomposition was packed within approximately 2 h using 512 cores on 16 nodes. The data set with the 0.6 mm decomposition was packed in approximately 3 h using 256 cores on 16 nodes. Although above stated differently, this core reduction per node was done because of a memory problem which again resulted from the not optimal implementation of the serialized header communication and combination.

As an overal result it can be stated, that we are quite satisfied with the direction the performance results of the current implementation are pointing at. Especially if one extrapolates the numbers given in Table 2 for the tar algorithm to data sets of sizes as the ones given in Table 3. It would never be feasible to pack the full 0.6 mm data set with tar within approximately 3 weeks.

5 Summary and Outlook

In the first part of this work it was shown that the handling of fine grain data sets, as they are produced by parametric studies or other workflow like mechanisms executed on high performance computers, can not be done in a convenient way by serial applications. This was demonstrated by the application of tar to data sets of a representative fine grain structure.

In the second part a storage and indexing concept was proposed which is supposed to solve the addressed problems. Afterwards the initial implementation of the indexing scheme by means of FORTRAN 2003 as well as the realization of MPI-parallel indexing and packaging application is described.

In the results section the performance evaluation of the implemented algorithm is shown for its application to representative data sets as well as the results of the application to real life data originating from a use case presented in the first part.

Although the performance reached by the current implementation is already in reasonable regions, e.g. the application of the algorithm to a representative data set with 1.6 MB average file size was executed on 256 cores with over 50 % efficiency per core, some aspects of the application need improvement.

The first point to focus on is the usage of a final communication step which has been noticed to be very inefficient. An improvement would be to send the serialized indicees right after the indexing step to the first process. With that process not involved in indexing and packaging it can collect, combine and store the global index while its partners do the data storage.

The second point which is supposed to result in a performance gain especially for fine grain data sets with an average file size way below the per core memory amount is the aggregation of data during the storage phase for collective write operations. Before this strategy is implemented, although it has in other projects [4] already been proven to be effective, it should of course be finally understood due to what effect the performance penalty for data sets with bigger average file size is induced.

The third point to be analyzed regards the storage scheme of the data. Here the imitation of a well known storage scheme like ustar[6] or the usage of the HDF5[7] library should be considered if recovery of the archive from index or archive corruption is an issue.

References

1. Rauhut, G., Schweiger, S., in *High Performance Computing in Science and Engineering' 04*, ed. by Krause, E., Jäger, W., Resch, M. Potential Energy Surfaces of Unusual Double Proton Transfer Reactions, (Springer Berlin Heidelberg 2005), pp. 323–331, doi: 10.1007/3-540-26589-9_30
2. Schneider, R,. Hindenlang, U., Resch, M.: Identification of Anisotropic Elastic Material Properties from Micro-Fem Simulations for Natural Materials. inSiDE, **7** No. 2 pp. 12–19, (2009)
3. Shipman, G., M., et al.: Lessons Learned in Deploying the World's Largest Scale Lustre File System, Proceedings of the Cray User Goup conference (CUG 2010), Edinburgh, Great Britain,

[6]http://ieeexplore.ieee.org/xpls/abs_all.jsp?arnumber=4039999&tag=1

[7]http://www.hdfgroup.org/HDF5/

May 24–27 2010 https://cug.org/5-publications/proceedings_attendee_lists/CUG10CD/pages/11-program/final_program/CUG10_Proceedings/pages/authors/11-15Wednesday/13A-Shipman-paper.pdf
4. Močnik, J., Novak, M., Focht, E., in *High Performance Computing on Vector Systems 2011*, ed by Resch, M., et al. I/O Forwarding for Quiet Clusters, (Springer Berlin Heidelberg 2011), pp. 21–39, doi: 10.1007/978-3-642-22244-3_2

Part III
Computational Engineering Applications and Multi-Physics Simulations

Direct Numerical Simulations of Film Cooling in a Supersonic Boundary-Layer Flow on Massively-Parallel Supercomputers

Michael Keller and Markus J. Kloker

Abstract Future rocket-nozzle extensions have to be thermally protected by a film of cooling gas. Here, the cooling film is generated by wall-parallel cooling-gas injection through a backward facing step. In a first step, a generic laminar flat-plate boundary-layer flow with external Mach number 2.6 and zero streamwise pressure gradient is used, where air is employed as hot and cooling gas. Direct numerical simulations are performed allowing for the reliable detection of any enhanced laminar-flow instability. Using compact finite differences or compact data filtering, tridiagonal sets of equations have to be solved employing the pipelined Thomas algorithm in order to compute various spatial derivatives or low-pass filtered data. In contrast to the NEC-SX8/9 vector machines with few, powerful compute nodes the solution of this tridiagonal systems turned out to be a major bottleneck on the massively parallel Cray-XE6 system. In order to avoid processor idling fully explicit and sub-domain compact finite differences are implemented and applied to the wall-parallel cooling-gas injection problem. The numerical results and performance data on the CRAY-XE6 system are compared to the regular, globally compact finite-difference scheme.

1 Introduction

Typically, a gain in power output of rocket engines is achieved by increasing thrust-chamber pressure and temperature resulting in heat loads that exceed the temperature limits of today's available materials. According to Haidn [6] power levels of up to 30 GW are achieved for solid rocket engines and up to 20 GW for

M. Keller (✉) · M.J. Kloker
Institute of Aerodynamics and Gas Dynamics, Pfaffenwaldring 21, 70550 Stuttgart, Germany
e-mail: keller@iag.uni-stuttgart.de; kloker@iag.uni-stuttgart.de

M.M. Resch et al. (eds.), *Sustained Simulation Performance 2013*,
DOI 10.1007/978-3-319-01439-5_8,

liquid rocket engines. Hence, in addition to the ongoing progress in the field of material science innovative and efficient cooling strategies have to be developed in order to protect the thermally highly loaded regions of combustion chambers and rocket nozzles of next-generation space-transportation systems from thermal failure. For instance, the nozzle extension of Vulcain 2, ESA's Ariane 5 main engine, is protected by several cooling systems (Winterfeldt et al. [19]). Regenerative and dump cooling systems are used, where the wall is cooled by convection. In addition, film-cooling is applied by wall-parallel injection of the collected turbine-exhaust gas through a backward-facing-step ring. Other cooling techniques used in rocket engines are ablative cooling, which is often found in solid rocket boosters, and radiative cooling (Haidn [6]). In general, film cooling is suited for active thermal protection systems, since it allows for an adjustment during flight and, thus, is also applicable to external-surface cooling of super- and hypersonic vehicles during reentry or cruise. Note that throughout literature the terminology with respect to film cooling is inconsistent. Here, we are referring to the procedure where a gaseous cooling film is established by wall-normal or wall-parallel injection through discrete holes or slits (the first is also called effusion cooling). Sometimes, this is also referred to as curtain cooling, because a thin film is covering the surface. Please note that surfaces, which are wetted with a liquid (propellant) that evaporates and, thus, provides cooling are not considered.

The cooling film can be generated by wall-parallel injection through a backward facing step (e.g. Aupoix et al. [1], Juhany and Hunt [9], and Konopka et al. [11]), through porous materials (e.g. Gühlhan and Braun [5]), or through discrete holes and spanwise slits. The latter has been investigated, among others, by Linn and Kloker, see [13–15]. Direct numerical simulations of effusion cooling in laminar super- and hypersonic boundary layers were used to analyze the effects of geometric and gas-dynamic parameters, such as hole spacing, hole arrangement, Mach number, Reynolds number, blowing rate, inclination angle, compound angle, and wall-temperature condition. From a cooling-performance point of view, slits are better than holes when imposing the same total mass flux, and staggered hole rows are better than aligned rows. Also, a small spanwise spacing is preferable in order to generate a more homogeneous cooling film. It was furthermore demonstrated that an inclination angle and compound angle as well as the wall-temperature condition – isothermal, adiabatic or radiative-adiabatic – strongly affect the cooling performance. The results for an isothermal, cooled wall (short-duration shock-wave experiments) differ from the results with a radiative-adiabatic wall (flight condition) and a transfer of the results is therefore difficult. Experimental studies have been performed by Heufer and Olivier, see e.g. [7, 8]. They investigated the effects of spanwise slits and discrete holes as well as the influence of blowing rate, Mach number, Reynolds number, various cooling gases, and flow acceleration. A comparison of the numerical results of Linn and Kloker with the experimental findings of Heufer and Olivier for a laminar Mach-2.67 boundary layer with an isothermal, cool wall showed good agreement. Numerical investigations have also been performed by Dahmen et al., e.g. [4], using a multi-resolution finite-volume solver in order to develop multi-scale techniques for highly resolved vortical structures.

The applied numerical method is based on a high-order compact finite difference scheme [2, 3, 10, 12]. The use of compact finite differences requires the solution of a tridiagonal set of equations. This is done by the Thomas algorithm [2, 3]. However, the spatial coupling of each derivative inhibits a parallelization and pipelining is commonly used [16]. Yet, moving the DNS code from a vector computer with few, powerful compute nodes to a massively-parallel supercomputer with thousands of processors shows that pipelining soon reaches its limit. Thus, other methods have to be developed and implemented allowing for an "independent solution" of the tridiagonal set of equations. Two approaches are presented: First, a sub-domain compact finite-difference scheme is introduced, where explicit finite differences in several fashions are used at the sub-domain boundaries to compute the exterior derivatives of the pure compact finite differences. Secondly, inside the sub-domain explicit finite differences of up to 8th-order are implemented in order to drop the solution of a tridiagonal set of equations. The explicit, the sub-domain compact, and the globally compact finite-difference schemes are applied to the numerical simulation of the film-cooling problem with wall-parallel cooling-gas injection.

The paper is organized as follows: Sect. 2 describes the governing equations and the numerical method. It contains an overview of the spatial discretization for the explicit, sub-domain compact, and globally compact finite-difference scheme. The numerical results for the wall-parallel cooling-gas injection into a supersonic laminar boundary-layer flow are presented in Sect. 3. Finally, Sect. 4 summarizes the main findings and contains some concluding remarks.

2 Numerical Method

A time-accurate direct numerical simulation (DNS) is used to compute the flow quantities. DNS solve the governing equations without turbulence modeling and allow for reliable detection of any enhanced laminar-flow instability leading to self-excited unsteadiness (by grown numerical background noise). The present simulations are carried out on block-structured Cartesian grids.

2.1 Governing Equations

The governing equations are the continuity equation, the three-dimensional compressible Navier-Stokes equations and the energy equation, in the following written in dimensionless vectorized form:

$$\frac{\partial \rho}{\partial t} + \nabla \cdot (\rho \boldsymbol{v}) = 0 \quad , \tag{1}$$

$$\frac{\partial (\rho \boldsymbol{v})}{\partial t} + \nabla \cdot (\rho \boldsymbol{v}\boldsymbol{v}) + \nabla p = \frac{1}{Re_\infty} \cdot \nabla \boldsymbol{\sigma} \quad \text{and} \tag{2}$$

$$\frac{\partial (\rho e)}{\partial t} + \nabla \cdot (p + \rho e)\, \boldsymbol{v} = \frac{1}{(\kappa - 1)\, Pr\, Re_\infty\, M_\infty^2} \nabla \cdot (\vartheta \nabla T) + \frac{1}{Re_\infty} \nabla \cdot (\boldsymbol{\sigma} \boldsymbol{v}) \, , \tag{3}$$

where

$$\boldsymbol{\sigma} = \mu \left(\left(\nabla \boldsymbol{v} + \nabla \boldsymbol{v}^T \right) - \frac{2}{3} (\nabla \cdot \boldsymbol{v})\, \mathbf{I} \right) \tag{4}$$

describes the viscous stresses, and

$$e = c_v \cdot T + \frac{1}{2} \left(u^2 + v^2 + w^2 \right) \tag{5}$$

is the internal energy per mass unit. The governing equations are non-dimensionalized with the reference length $L^\star = (\mu_\infty^\star \cdot Re_\infty)/(\rho_\infty^\star \cdot u_\infty^\star)$ and the freestream values of velocity, density, temperature, viscosity and conductivity at the inflow. Note that the pressure is non-dimensionalized by $\left(\rho_\infty^\star {u_\infty^\star}^2\right)$. Air is treated as a non-reacting, calorically perfect gas with constant Prandtl number $Pr = 0.71$ and constant specific-heat ratio $\kappa = c_p/c_v = 1.4$. The set of equations is closed by the equation of state $p = (\rho T)/(\kappa\ M_\infty^2)$. Sutherland's law is used to calculate the dynamic viscosity μ as a function of temperature [18]. In these equations, $\boldsymbol{v} = (u, v, w)^T$ represents the velocity vector, p labels the pressure, ρ indicates the density, T is the temperature, and t denotes the time. The non-dimensional parameters are the Mach number M_∞, the Prandtl number Pr and the Reynolds number Re_∞. The latter is defined as $Re_\infty = \left(\rho_\infty^\star \cdot u_\infty^\star \cdot L^\star\right) / \left(\mu_\infty^\star\right)$ and set to $Re_\infty = 10^5$. $\mathbf{I}$ is the identity matrix. The subscript $_\infty$ refers to freestream values and the asterisk * labels dimensional quantities. The governing equations are solved using the in-house Fortran code NS3D [2, 3].

2.2 *Spatial Discretization and Time Integration*

By default, the code uses compact finite differences of up to 6th-order for the spatial discretization in streamwise and wall-normal direction [2, 3]. This requires the solution of a tridiagonal set of equations, which is carried out by the Thomas algorithm. This algorithm is very efficient and essentially consists of a forward loop and a backward loop in order to obtain the solution of the linear system of equations. Blocking MPI communication is needed, since the backward-loop computation

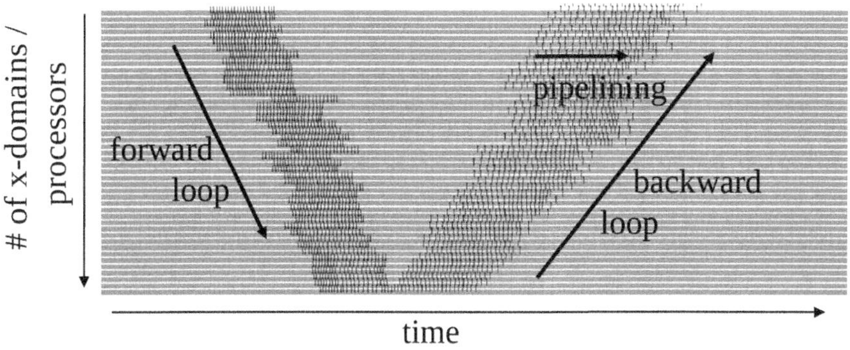

Fig. 1 MPI communication after one Runge-Kutta sub-step (*black lines*) for a test case with 64 MPI tasks in streamwise direction

depends on the results of the forward loop. In order to reduce the idle time, the algorithm is pipelined. Up to 21 or 25 derivatives have to be computed in streamwise and wall-normal direction, respectively. Thus, after the first-loop computation of the first derivative CPU 1 starts working on the first-loop computation of the subsequent derivative, instead of waiting for the results of the backward loop. The pipelining technique is a valuable means and perfectly suited for vector machines with few, powerful compute nodes. However, it is not efficient enough if the system of equations is distributed over a large number of CPUs. Figure 1 illustrates the MPI communication after one Runge-Kutta sub-step on the massively parallel CRAY-XE6 system for a test case with 64 MPI tasks in streamwise direction. The communication is visualized by the black lines using the CRAY-Apprentice^2 performance-analysis tool. Despite pipelining the processors spend a lot of time idling, since they have to wait for the results of the backward loop.

In order to decrease the dead time and properly harvest the potential of a massively parallel system the tridiagonal set of equations has to be split up and solved independently by each MPI task. Hence, a sub-domain compact scheme is developed and implemented into the code. In addition, explicit finite differences of up to 8th-order are implemented. The latter even avoids the solution of the tridiagonal set of equations at all, but has some shortcomings that are demonstrated later.

Using the established 6th-order compact scheme the first and second derivatives (ϕ'_j, ϕ''_j) of an arbitrary flow variable ϕ at grid point j are computed by the following equation:

$$\alpha\left(\phi'_{j-1}, \phi''_{j-1}\right) + \beta\left(\phi'_j, \phi''_j\right) + \gamma\left(\phi'_{j+1}, \phi''_{j+1}\right) = \frac{1}{q}\left(a\phi_{j-2} + b\phi_{j-1} + c\phi_j + d\phi_{j+1} + e\phi_{j+2}\right). \tag{6}$$

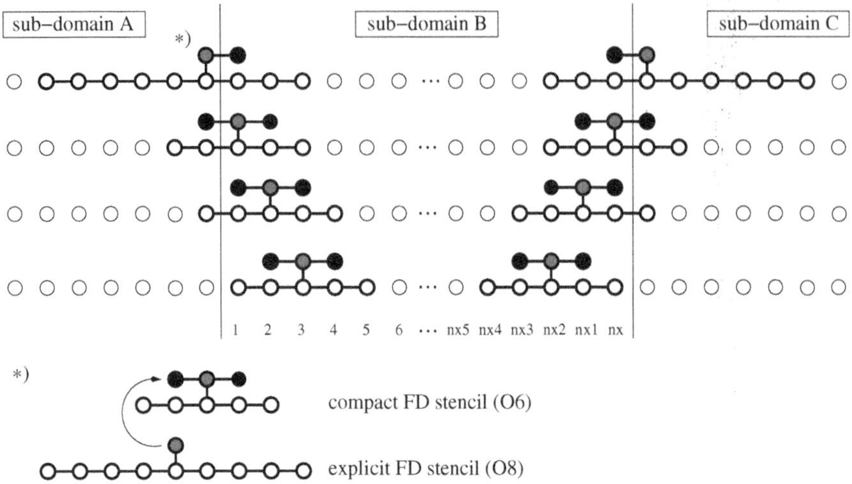

Fig. 2 Finite-difference stencils for the sub-domain compact scheme using one ghost point at each sub-domain boundary ($j = nx$ of sub-domain A and $j = 1$ of sub-domain C). The *black solid circles* mark the derivatives ϕ' and ϕ'', the *white circles* indicate the functional values ϕ

Applying explicit finite differences of 8th-order the first and second derivatives are given by:

$$\phi'_j, \phi''_j = \frac{1}{q}\left(a\phi_{j-4} + b\phi_{j-3} + c\phi_{j-2} + d\phi_{j-1} + e\phi_j + f\phi_{j+1} + g\phi_{j+2} + h\phi_{j+3} + i\phi_{j+4}\right). \tag{7}$$

Due the higher order a larger stencil is required. The minor diagonals and the coupling have vanished. Near *physical* domain boundaries one-sided and biased stencils of reduced order ($\geq$4) are used for both schemes (not shown).

The numerical procedure of the sub-domain compact scheme is illustrated in Fig. 2. Compact finite differences of 6th-order are used within the *full* sub-domain. Out of the sub-domain boundaries explicit finite differences of 8th-order are employed replacing the outermost derivative of the compact finite-difference stencil. As shown at the bottom of Fig. 2 this results in a one-sided compact finite-difference stencil. In order to compute virtually every derivative $j = 1$ to $j = nx$ by a central compact finite difference at least one ghost point ($j = nx$ of sub-domain A and $j = 1$ of sub-domain C) is needed. It is computed by the special, biased compact finite-difference indicated by the asterisk. In contrast to the regular compact finite-difference scheme the tridiagonal set of equations can now be computed separately for each MPI task. Using one ghost point is the default setting in the NS3D code. However, an option with zero and two ghost points, respectively, is also tested.

Of course, the mathematical properties of the sub-domain compact versions are not identical to the globally compact scheme. The question is how far the derivatives palpably couple, and thus how significant the suppression of full coupling is, see below. Note that the physical domain boundaries are treated identically to the globally compact scheme.

According to [2, 10] an analysis of the modified wavenumber allows for an accurate evaluation of the numerical properties of the finite differences in a model setting. For that purpose a spatially periodic wave given by $\phi = e^{ikx}$ is considered. In this equation $i = \sqrt{-1}$ is the imaginary unit and k, which is real, represents the wavenumber. The analytically obtained first derivative is $(\partial\phi/\partial x)_{ana} = \phi'_{ana} = ik\,\phi$, and the second derivative reads $(\partial^2\phi/\partial x^2)_{ana} = \phi''_{ana} = -k^2\,\phi$. Introducing $k^\star = k\Delta x$ yields:

$$k^\star = -i\Delta x \frac{\phi'_{ana}}{\phi} \quad \text{and} \quad k^{\star 2} = -\Delta x^2 \frac{\phi''_{ana}}{\phi}\,. \tag{8}$$

Using the numerically determined derivatives of Eqs. 6 and 7 results in the definition of the modified wavenumber:

$$k^\star_{mod} = -i\Delta x \frac{\phi'_{num}}{\phi} \quad \text{and} \quad k^{\star 2}_{mod} = -\Delta x^2 \frac{\phi''_{num}}{\phi}\,; \tag{9}$$

$k^* = 0$ corresponds to a wave with an infinitesimal fine grid resolution $n_w = \infty$. Here, $n_w = 2\pi/k^\star$ indicates the number of grid points per wave length. $k^* = \pi$ represents the least resolved wave with two points per wave length (wiggle mode, $n_w = 2$). The modified wavenumber is complex for non-symmetric finite-difference stencils. A deviation in the real part represents an amplitude error in the first derivative, whereas the imaginary part causes a phase shift in the first derivative. Recall that the latter results in an *amplitude* error in the solution of the (convective) transport-equation part of the flow equations, which is zero with symmetric (central) stencils, because the imaginary part of the wavenumber is zero.

Symmetric finite-difference stencils are used to compute the first derivatives of the viscous terms and the second derivatives. Note that a Laplace formulation is used for the computation of the viscous terms, i.e. second derivatives are used where possible. The corresponding coefficients of Eqs. 6 and 7 are given in Tables 1 and 2 indicated by the label D0. Numerical damping for high $k^\star$ is needed for the computation of the first derivatives of the convective terms. To that end non-symmetric, alternating finite differences are used [10]. Three different schemes with varying damping properties are implemented and labeled by D1, D2, D3, where D1 provides the least and D3 the highest damping, respectively. The coefficients for the forward-biased stencils are given in Tables 1 and 2. The backward-biased stencils follow by reversing the order of the right-hand side coefficients, and changing the sign: a (backward) $= -e$ (forward) in Table 1, for example [10].

The damping properties of the alternating back- and forward biased explicit finite differences are chosen such that they have similar damping properties as

Table 1 Finite-difference coefficients of Eq. 6 for the first and second derivatives using the 6th-order compact scheme after [3]

	D0 (ϕ')	D1 (ϕ')	D2 (ϕ')	D3 (ϕ')	D0 (ϕ'')
α	1/5	1/5	1/5	1/5	2/15
β	3/5	3/5	3/5	3/5	11/15
γ	1/5	1/5	1/5	1/5	2/15
a	−1	0	2	7	3
b	−28	−9	−9	−18	48
c	0	−11	−15	−36	−102
d	28	19	19	38	48
e	1	1	3	9	3
q	60 h	30 h	30 h	30 h	60 h^2
Order	O6	O6	O6	O6	O6

Table 2 Finite-difference coefficients of Eq. 7 for the first and second derivatives using the 8th-order explicit scheme

	D0 (ϕ')	D1 (ϕ')	D2 (ϕ')	D3 (ϕ')	D0 (ϕ'')
a	3	−7.335153682	−2.741751740	0.703299239	−9
b	−32	29.775732900	2.319851354	−18.272059460	128
c	168	−41.075603620	51.846886800	121.538754800	−1,008
d	−672	−33.775732900	−6.319851354	14.272059460	8,064
e	0	−961.178485400	−1,156.210270000	−1,302.484108000	−14,350
f	672	1,310.224267100	1,337.680149000	1,358.272059000	8,064
g	−168	−377.075603620	−284.153113200	−214.461245200	−1,008
h	32	93.775732900	66.319851350	45.727940540	128
i	−3	−13.335153682	−8.741751740	−5.296700761	9
q	840 h	840 h	840 h	840 h	5,040 h^2
Order	O8	O8	O8	O8	O8

the respective compact finite differences ($Im(k^\star_{mod,expl}) = Im(k^\star_{mod,comp})$ at $k^\star = 0.5,\ 1.0,\ 1.5$ and π, see [3, 10]). Figure 3a shows the real part of the modified wavenumber for the explicit and compact finite differences. Despite its higher order, the explicit scheme deviates earlier from the exact solution. Compared to the compact scheme, the explicit scheme is slightly less robust and a higher grid resolution is required for badly resolved waves. Figure 3 also contains the biased compact stencil used for the computation of the first derivatives at the ghost points. Due to the explicit influence, the aliasing limit ($k^\star$ at $Re(k^\star_{mod}) = max$, Fig. 3a) is slightly decreased. The imaginary part of the modified wavenumber is given in Fig. 3b. It illustrates the properties of the D1, D2 and D3 scheme, and of the biased stencil for the outermost ghost points. The latter has a small (unwanted) imaginary part only.

The explicit, the sub-domain compact, and the globally compact scheme are used to compute the derivative of a test function $\phi(x) = sin(kx)$ with $k = 1, 2, 4$ and 8. This corresponds to waves with a grid resolution of $n_w = 30,\ 15,\ 7.5$ and 3.75 ($k^\star \simeq 0.209,\ 0.419,\ 0.838$ and 1.676), respectively, on sub-domains with lengths

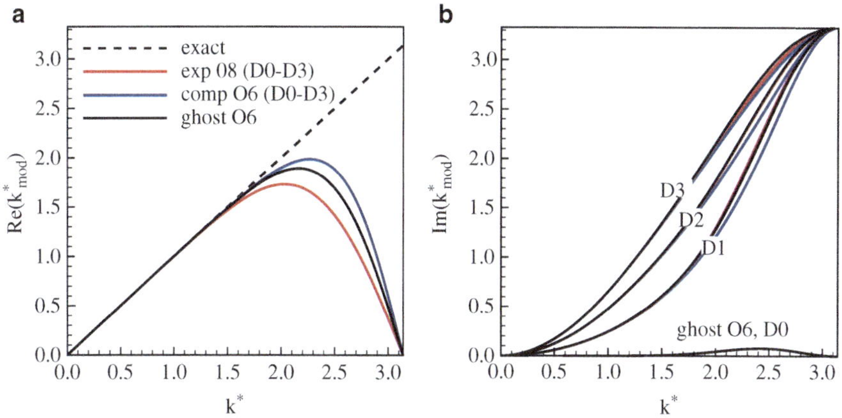

Fig. 3 Real part (**a**) and imaginary part (**b**) of the modified wavenumber for the explicit and compact schemes. The one-sided compact finite difference stencil used for the computation of the derivatives at the ghost points is also included and indicated by the *black solid lines*

of 2π and 30 points each. The analytical derivative reads $\phi'_{ana}(x) = k\ cos(kx)$. Figure 4 illustrates the numerical error $(\phi'_{num}(x) - \phi'_{ana}(x))$. Note that the function $(\phi'_{num}(x) - \phi'_{ana}(x))$ inherits the corresponding wavenumber of $\phi(x)$, since $\phi'_{num}(x) = Re(k_{mod})\ cos(kx) - Im(k_{mod})\ sin(kx) = A\ cos(kx - \varphi)$ and $\phi'_{ana}(x) = k\ cos(kx)$; hence $(\phi'_{num} - \phi'_{ana}) \sim cos(kx - \varphi)$. For centered finite differences k_{mod} is real and it holds $(\phi'_{num} - \phi'_{ana}) = (q - 1)k\ cos(kx)$, with $q = k_{mod}/k \leq 1$. Due to the higher order of the explicit scheme a smaller error is observed for the well resolved waves ($k^\star \simeq 0.209,\ 0.419$), with the relative error being about 10^{-8} or 10^{-6}, respectively. However, as the number of points per wave length is reduced the error of the explicit scheme is larger than the error of the (sub-domain) compact scheme, which later deviates from the exact solution (cf. Fig. 3a). Using the sub-domain compact scheme an additional error at the sub-domain boundaries is introduced. However, this additional error is remarkably small. As mentioned above three different versions are implemented, where two, one or zero ghost points are used at the sub-domain boundaries. Compared to the regular, globally compact finite-difference scheme, the sub-domain compact scheme without ghost points shows expectedly the largest additional error (orange line). It can be significantly reduced by using ghost points for the derivatives. This marginally increases the computational cost, but results in a significant weakening of the explicit influence at the sub-domain boundaries. Thus, the version with one ghost point is chosen as the default setting in the NS3D code.

For 3-d simulations the derivatives in spanwise direction can be computed by means of a Fourier-spectral discretization, due to the assumed periodicity of the flow field. Alternatively, compact finite differences of 6th-order can be used. The classical explicit 4th-order Runge-Kutta procedure is applied for the integration in time. The fundamentals of the code can be found in [2, 3, 10, 14].

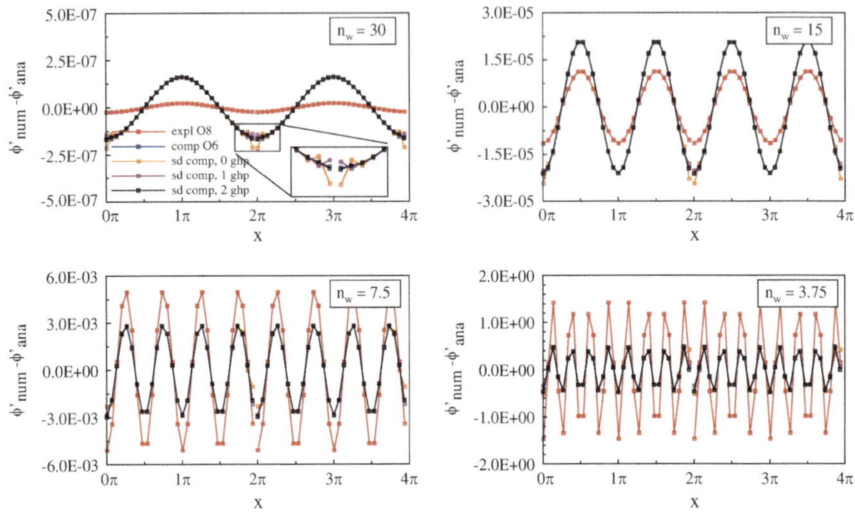

Fig. 4 Numerical error $(\phi'_{num}(x) - \phi'_{ana}(x))$; $\phi(x) = sin(kx)$ with $k = 1, 2, 4$ and 8

2.3 *Computational Domain, Boundary Conditions and Initial Condition*

The wall-parallel cooling-gas injection is realized by means of a splitter plate, separating a boundary-layer flow from a cool Poiseuille flow, see Fig. 5a. The 2-d simulations are performed for a generic laminar boundary-layer flow over a flat plate with zero pressure gradient. The freestream Mach number, temperature and pressure are given by $M_\infty = 2.6$, $T_\infty = 500\,\mathrm{K}$ and $p_\infty = 0.14\,\mathrm{bar}$, respectively. These values are consistent with the experiments conducted at the shock wave laboratory of RWTH Aachen University. The experiments are performed for a turbulent boundary-layer flow, being in progress in the numerical work not shown here. Figure 5a illustrates a sketch of the computational domain. Note that the spatial resolution (Fig. 5b) has been chosen extremely fine, neither requested by the flow physics nor the highly accurate code: A benchmark simulation, from the viewport of parallelization was to be generated.

A self-similarity solution of boundary-layer theory provides the flow variables prescribed at the boundary-layer inflow ($x_0/D = -12.9$, $y/D > 0$). Note that all length scales are presented in terms of the slit diameter $D = 2\,\mathrm{mm}$. At the outflow, $x_N/D = 27.3$, all flow quantities are computed using a 2nd-order extrapolation. Due to the supersonic flow regime, a special damping zone to suppress reflections is not needed. At the wall ($y/D = 0$ and $y/D = -1$), the no-slip, no-penetration boundary conditions are imposed on the velocity components. The pressure is calculated according to $\partial p/\partial y|_w = 0$ and the density is computed from the equation of state. The wall is assumed to be isothermal with $T_w = 293\,\mathrm{K}$ (short-duration shock-wave experiments). Note that thermal conduction within the wall is neglected

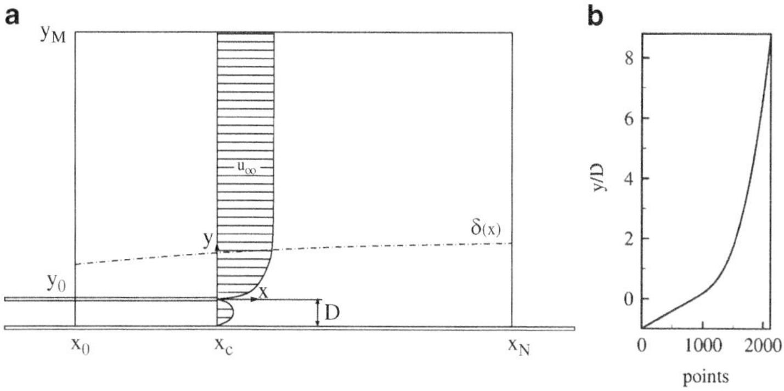

Fig. 5 Computational domain (**a**) and grid stretching in wall-normal direction (**b**)

in this study. At the freestream boundary (y_M), all flow variables are computed such that the gradient along spatial characteristics is zero, except for the pressure, which is computed from the equation of state. In wall-normal direction the computational domain extends to a height of $y_M/D = 8.8$. This corresponds to approximately 30 boundary layer thicknesses δ_{99} (taken at the inflow). The grid is equidistant in streamwise direction. In wall-normal direction the grid is equidistant within the channel ($-1 < y/D < 0$), for $y/D > 0$ a 3rd-order polynomial grid stretching is used (Fig. 5b). The origin of the coordinate system is placed at the plate's end, which corresponds to a distance from the leading edge of 85.0 mm.

The cooling channel has a length of 12.9 D and a height of $D = 2$ mm. At its inflow ($x_0/D = -12.9$, $-1 < y/D < 0$) the pressure is set to $p_c/p_\infty = 1.1$ (case A), $p_c/p_\infty = 2.2$ (case B) and $p_c/p_\infty = 3.3$ (case C), respectively. The temperature is given by $T_c = 293$ K. The wall-normal velocity component is assumed to be zero, whereas the streamwise velocity component is extrapolated from the downstream interior (subsonic flow). Finally, the density is computed from the equation of state. Note that $M = 1$ cannot be exceeded within the channel, since u is extrapolated and no throat is used ($D = const$). The Reynolds number of the channel flow is $Re_c = (\rho_c \cdot u_c \cdot D)/(\mu_c) \approx 5{,}140$ for case A, $Re_c \approx 11{,}640$ for case B, and $Re_c \approx 17{,}840$ for case C, respectively, where the subscript c indicates the area-averaged values at the cooling-channel inflow. The thickness of the splitter plate corresponds to one step size in wall-normal direction Δy. The setup is a tentative ansatz for the investigation of wall-parallel cooling-gas injection.

The number of grid points used is $5{,}000(x) \times 2{,}125(y) = 10.625\ 10^6$. The domain is disjoint in 2,500 blocks with $50(x) \times 85(y) = 4{,}250$ grid points each. The simulations are performed on the CRAY-XE6 system at the Federal High Performance Computing Center Stuttgart. The system provides a total of 3,552 compute nodes/113,664 cores resulting in a theoretical peak performance of about one PFlop/s. Two AMD Opteron processors (Interlagos) with 16 cores each represent one compute node. Using all 32 cores on one compute node yields a total number of 79 compute nodes for the simulations.

An overview of the simulation parameters is given in Table 3.

Table 3 Simulation parameters

Freestream Mach number M_∞	2.600 [–]	
Freestream temperature T_∞	1.000 [–]	500.0 K
Freestream pressure p_∞	0.106 [–]	0.140 bar
Wall temperature T_w	0.586 [–]	293.0 K
Slit diameter D	0.085 [–]	2.0 mm
Distance from the leading edge x_c	3.614 [–]	85.00 mm
Inlet pressure ratio p_c/p_∞ (case A)	1.1 [–]	0.145 bar
Inlet pressure ratio p_c/p_∞ (case B)	2.2 [–]	0.308 bar
Inlet pressure ratio p_c/p_∞ (case C)	3.3 [–]	0.462 bar
Cooling gas temperature T_c	0.586 [–]	293.0 K
$N_x \times N_y$	$5{,}000 \times 2{,}125$ [points]	
Δx	$0.68 \cdot 10^{-3}$ [–]	
$\Delta y_0 - \Delta y_M$	$0.10 \cdot 10^{-3} - 1.56 \cdot 10^{-3}$ [–]	
Δt	$0.11 \cdot 10^{-3}$ [–]	
$y_0 - y_M$	$-0.085 - 0.75$ [–]	
$x_0 - x_N$,	$2.525 - 5.94$ [–]	

2.4 *Parallelization*

The NS3D code uses a hybrid parallelization of both MPI (domain decomposition in streamwise and wall-normal direction) and OpenMP (spanwise direction, if applicable). A performance comparison of the explicit, sub-domain compact, and compact scheme is shown in Fig. 6. It illustrates the speedup and parallel efficiency, respectively, for a strong and weak-scaling test with 512 MPI tasks in streamwise direction. The term strong scaling refers to simulations, where the total domain size remains constant, whereas the domain size per processor is decreasing with increasing number of processors. For the weak scaling the total domain size increases, whereas the domain size per processor is kept constant. The speedup is defined as $SU = t_0/t_n$. Here, t_0 is the reference time for a run with 32 MPI tasks (one compute node) and t_n refers to the computational time using n MPI tasks. The parallel efficiency is given as $E = t_0/(n\ t_n)$. As mentioned above processor idling occurs, when compact finite differences are used, due to the use of the Thomas algorithm. For a large number of MPI tasks this results in significant performance losses. In contrast, using the sub-domain compact or the explicit scheme a high parallel efficiency is obtained. Due to cache effects a super-linear speedup is observed for the strong-scaling test.

For the present simulations (cf. Sect. 2.3) a specific CPU time (time per time step, grid point, and CPU) of 14.9 μs is measured when using the compact scheme. For the sub-domain compact scheme the specific CPU time reduces to 7.9 μs. Since no tridiagonal set of equations has to be solved for the explicit scheme the specific CPU time further reduces to 5.8 μs.

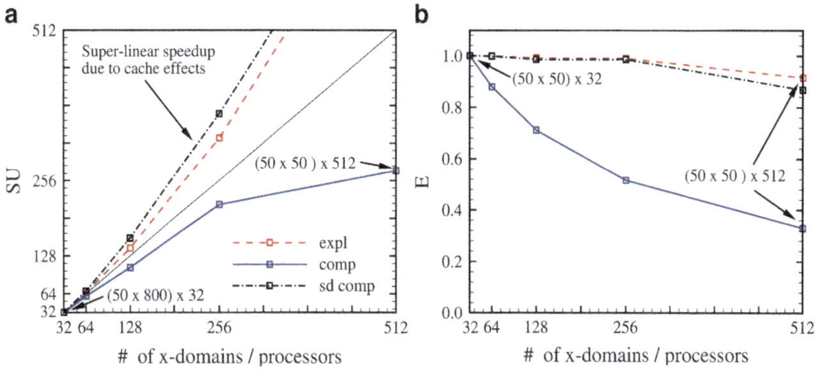

Fig. 6 Speedup of a strong-scaling test (**a**) and parallel efficiency of a weak-scaling test (**b**). The values in *brackets* indicate the number of grid points in x- and y-direction, respectively, for each sub-domain

3 Numerical Results

The explicit, sub-domain compact, and compact finite-difference schemes are applied to the wall-parallel cooling-gas injection problem. If not stated otherwise, the results of the sub-domain compact scheme are presented.

Figure 7 illustrates a Schlieren-type visualization of cases A–C using the computed density gradient $| \nabla \rho| = \sqrt{(\partial\rho/\partial x)^2 + (\partial\rho/\partial y)^2}$. The wall-normal velocity distribution is shown in Fig. 8. Due to the relatively low pressure at the channel inflow for case A, a weak expansion of the boundary-layer flow can be observed for $y/D > 0$. This is indicated by the slightly negative wall-normal velocity component. In contrast, cases B and C are characterized by a shock wave emanating from the trailing edge of the splitter plate (positive wall-normal velocity component) for $y/D > 0$. Due to the high pressure within the cooling channel, an expansion of the cooling gas occurs for $x/D > 0$ and $y/D < 0$, leading to a defection of the boundary-layer flow. Within the cooling-gas layer a wave train develops. The trailing-edge expansion wave is reflected at the lower wall and converted into an oblique shock wave. The latter is a again reflected at the shear-layer edge and converted into an expansion wave. The characteristic wave pattern depends on the channel-inflow pressure. For case C a self-excited unsteadiness develops, showing vortical structures in the shear-layer region.

A snapshot of the shear-layer vortices is shown in Fig. 9. They are visualized by means of numerical Schlieren and spanwise vorticity $\omega_z = (\partial v/\partial x) - (\partial u/\partial y)$. The vortices originate from the upper side of the splitter plate, where an unsteady recirculation region develops. Apparently, the vortices are a mixture of the early-stage mushroom-shaped Rayleigh-Taylor instability mainly based on density gradients, and the early-stage wave-like Kelvin-Helmholtz instability governed by shear, i.e. velocity gradients.

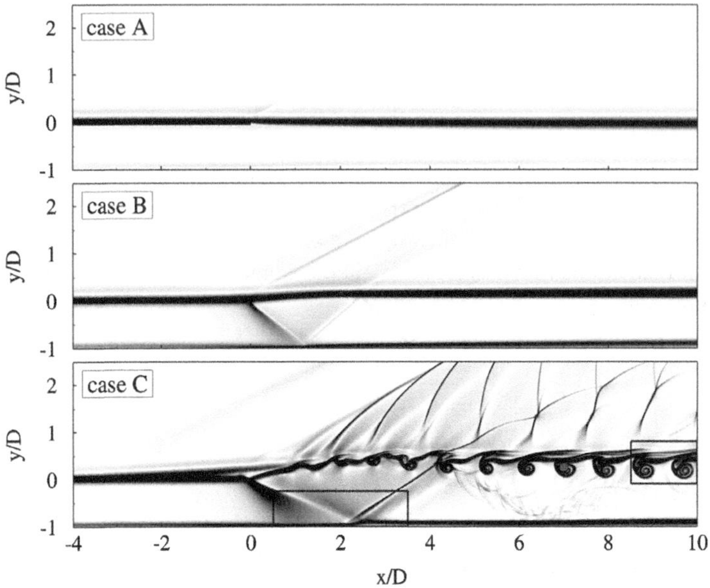

Fig. 7 Snapshot of the Schlieren-type visualization for cases A–C. The *insets* are illustrated in Figs. 9 and 11, respectively. The *red dot* indicates the position of the time signal presented in Fig. 10

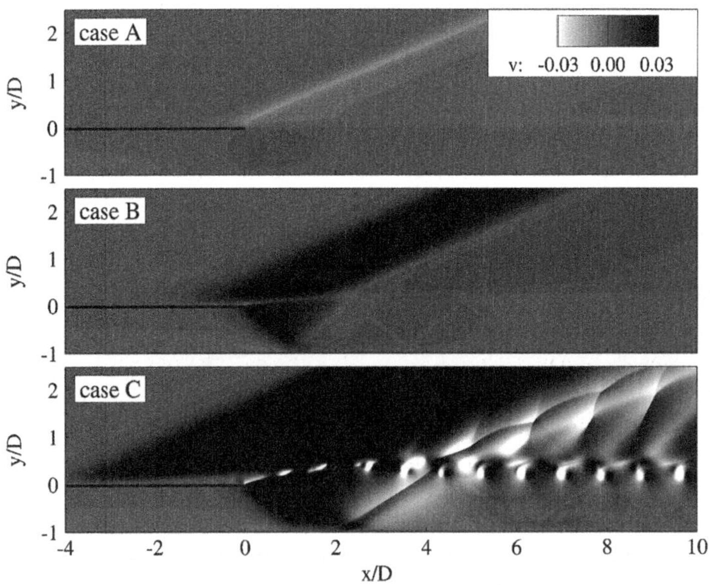

Fig. 8 Snapshot of the wall-normal velocity distribution for cases A–C. The *black solid line* indicates the splitter plate

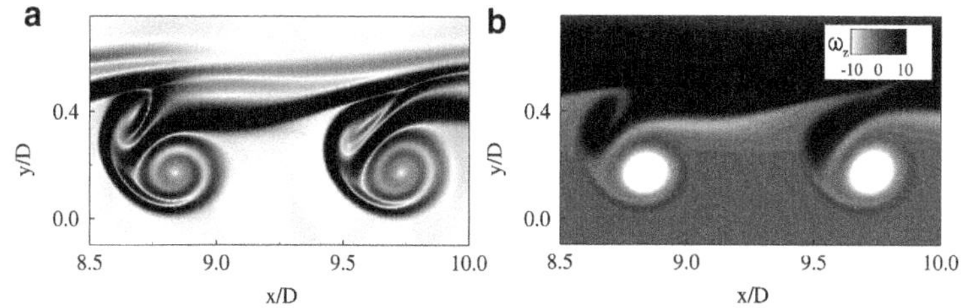

Fig. 9 Snapshot of the shear-layer vortices for case C visualized by means of numerical Schlieren (**a**) and spanwise vorticity (**b**)

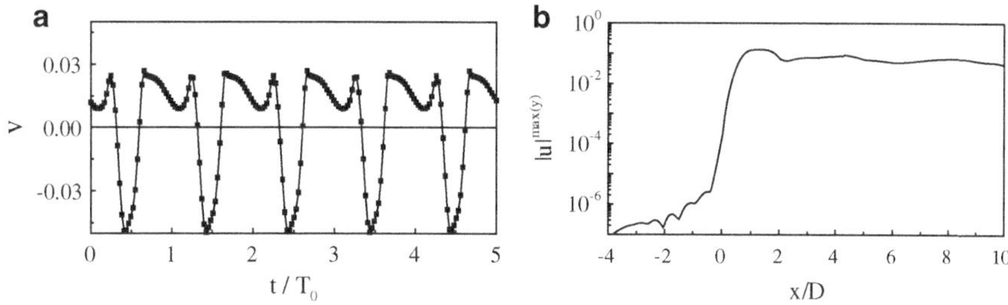

Fig. 10 (**a**) Temporal evolution of the wall-normal velocity component for case C at position $x/D = 7$ and $y/D = 0.25$ (cf. *red dot* in Fig. 7). (**b**) Downstream development of the maximum (over y) u-velocity for the vortex-shedding frequency of case C

Figure 10a shows the temporal evolution of the wall-normal velocity component at position $x/D = 7$ and $y/D = 0.25$. The position is indicated by the red dot in Fig. 7. A dominant frequency of 280 kHz can be identified for the vortex shedding. The streamwise development of the maximum u-velocity (over y) for this particular frequency is illustrated in Fig. 10b. A significant increase is observed at the trailing edge of the splitter plate indicating that the instability originates from this region.

The impingement region of the trailing-edge expansion wave for case C is presented in Fig. 11. It illustrates a snapshot of the small, unsteady separation bubble developing at the lower wall. The fluid is strongly accelerated and, thus, the gradients at the wall are increased. As the wall-normal resolution is decreased by a factor of 5, wiggles appear (Fig. 11b) without additional filtering. Using the explicit scheme the simulation even blows up, which indicates that it is slightly less robust than the (sub-domain) compact scheme, where the computation was completed successfully. Keep in mind that the boundary treatment is different with the explicit scheme: Two more points near the wall have to be computed by special finite-difference stencils without biasing and, thus, numerical damping.

The Mach number and temperature distribution for cases A–C is shown in Figs. 12 and 13. Due to the strong expansion wave emanating from the splitter-plate edge for cases B and C, the temperature of the cooling gas decreases, while simultaneously the Mach number increases for $x/D > 0$. This is also illustrated

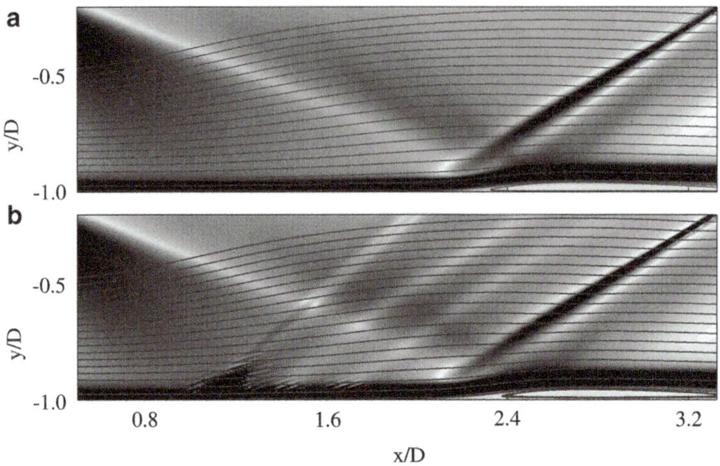

Fig. 11 Snapshot of the unsteady separation bubble for case C visualized by means of streamlines and numerical Schlieren. (**a**) Standard resolution in wall-normal direction Δy_{ref}, (**b**) $\Delta y = 5\Delta y_{ref}$

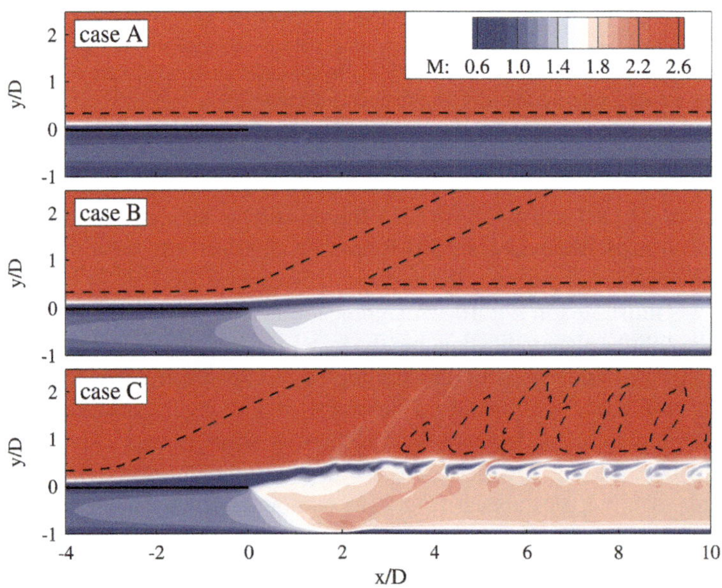

Fig. 12 Snapshot of the Mach number distribution for cases A–C. The *black solid line* indicates the splitter plate and the *black dashed line* represents the boundary layer thickness $\delta = 0.99u_\infty$

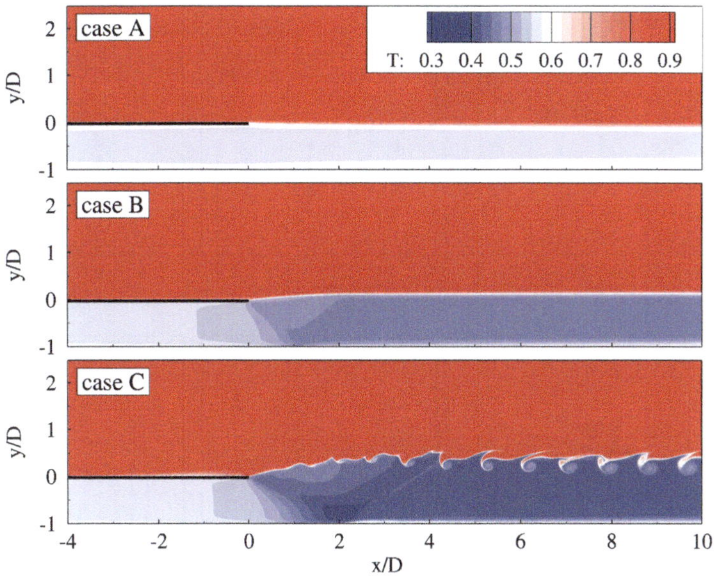

Fig. 13 Snapshot of the temperature distribution for cases A–C. The *black solid line* indicates the splitter plate

in Fig. 14, where the u-velocity, density and temperature profiles at $x/D = -4$ and $x/D = 10$ are shown, respectively. The profiles of case C are averaged in time over five fundamental periods. As mentioned above the Mach number cannot exceed unity within the channel with the used setup. Since the temperature is also nearly fixed (isothermal walls, constant cooling-gas temperature at the channel inflow), the maximum u-velocity within the cooling channel is approximately the same for all cases. However, the density increases resulting in an increasing mass flux as the channel-inflow pressure is increased. Figure 14 also contains the results for the explicit and compact scheme. All curves show a perfect agreement. Negligible differences can be observed for case C in the region of the unsteady shear-layer (position $x/D = 10$).

Figure 15 illustrates the *instantaneous* u-velocity, density and temperature profiles of case C at $x/D = 10$ for the explicit, sub-domain compact, and globally compact scheme, respectively. In contrast to the time-averaged profiles presented in Fig. 14 the local results within the unsteady shear layer can differ, especially between the explicit and the (sub-domain) compact scheme due to different phases at the sensing station caused by the differing uncontrolled numerical background noise serving as the seed for the developing structures. Note that the smooth distribution for $-0.8 \le y/D \le 0.3$ for the explicit scheme is a consequence of not cutting through a vortex. Globally, the results are the same as can be seen from Fig. 16.

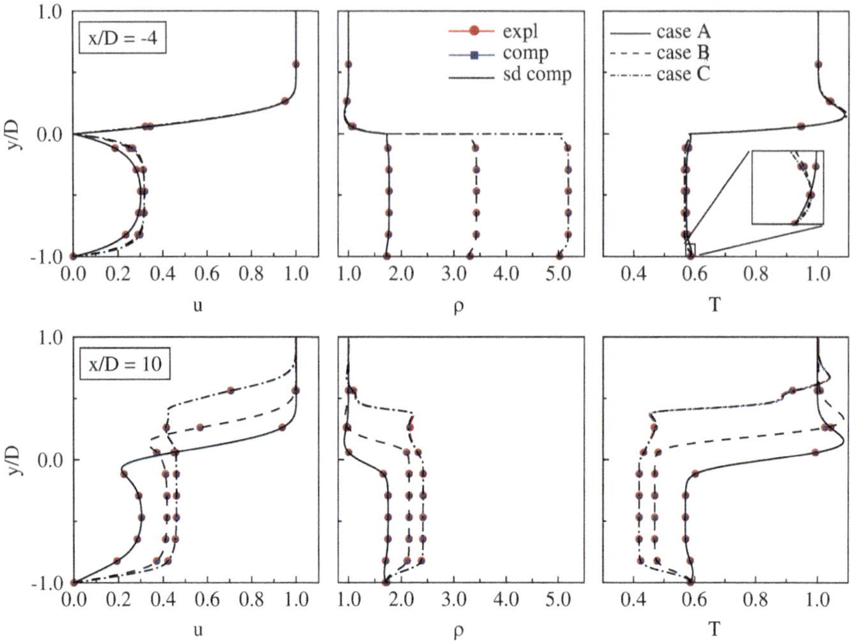

Fig. 14 u-velocity, density and temperature profiles at $x/D = -4$ (*top*) and $x/D = 10$ (*bottom*) for the explicit, sub-domain compact, and compact scheme. The results for case C are averaged in time over five fundamental periods

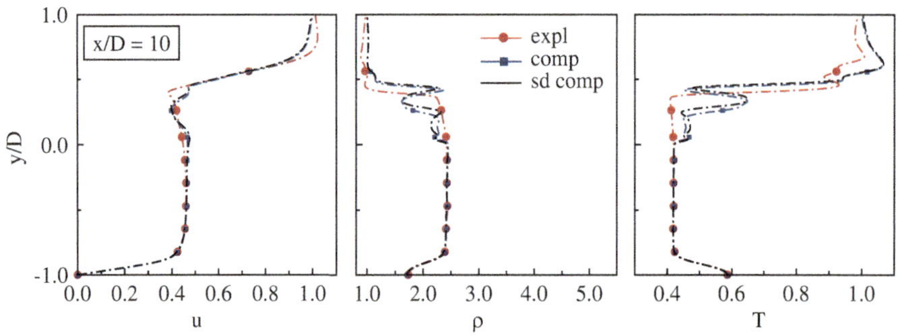

Fig. 15 Snapshot of the u-velocity, density and temperature profiles for case C at $x/D = 10$ for the explicit, sub-domain compact, and compact scheme

The downstream development of the pressure along the centerline of the cooling channel $y/D = -0.5$ is presented in Fig. 17a. Again, the results for case C are averaged in time over five fundamental periods. The figure illustrates that cases B and C are underexpanded, since the pressure at the end of the orifice is much higher than the pressure of the surrounding boundary-layer flow. For case A a

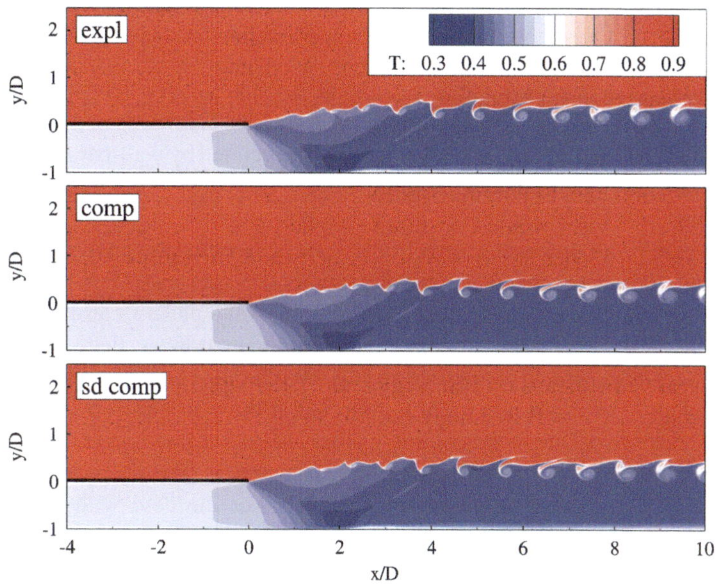

Fig. 16 Snapshot of the temperature distribution for case C for the explicit, sub-domain compact, and compact scheme

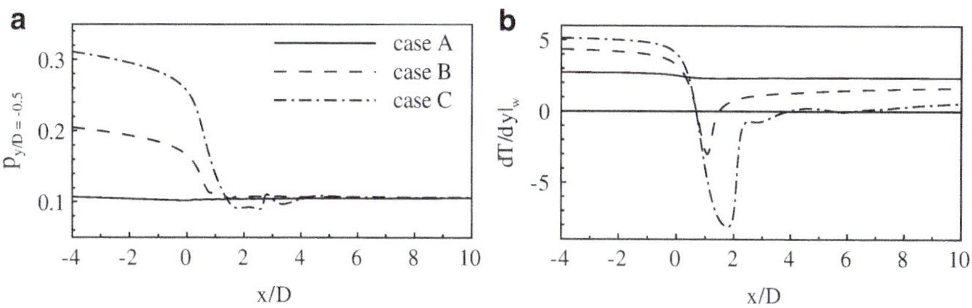

Fig. 17 Downstream development of the pressure along the channel centerline (**a**) and of the temperature gradient at the wall (**b**). The results for case C are averaged in time over five fundamental periods

matched flow condition is observed. The temperature gradient at the wall $(\partial T/\partial y)|_w$ is shown in Fig. 17b. The quantity is directly proportional to the wall-heat flux $\dot{q} = -\lambda(\partial T/\partial y)|_w$, where λ is constant for all cases, since isothermal walls are used. Negative values of $(\partial T/\partial y)|_w$ indicate a heat flux from the wall into the fluid, which means that wall is cooled. This is observed for cases B and C in the impingement region of the expansion wave. In general, increasing the channel-inflow pressure results in an underexpanded cooling-gas injection, which leads to a decrease of the cooling-gas temperature. Thus, the wall-heat flux decreases, which is beneficial in this respect.

4 Conclusions and Outlook

Film cooling in a generic laminar supersonic boundary-layer flow with external Mach number 2.6 and zero streamwise pressure gradient is investigated by means of direct numerical simulations. The cooling film is generated by wall-parallel cooling-gas injection through a backward facing step, where air is employed as hot and cooling gas.

Three cases with different channel-inflow pressure ratios are analyzed resulting in matched and underexpanded cooling-gas injection scenarios. For the underexpanded flow condition a characteristic supersonic wave pattern can be observed. Due to the expansion of the cooling gas, the main flow is deflected and an oblique shock is generated at the trailing edge. Within the cooling-gas layer a wave train develops. The expansion wave is reflected at the wall and converted into an oblique-shock wave. Due to the cooling-gas expansion, the cooling-gas temperature decreases significantly and the Mach number increases. The characteristic wave pattern depends on the channel-inflow pressure. For the case with the highest channel-inflow pressure ratio an undesirable, eventually film dissolving, self-excited unsteadiness develops, showing vortices in the shear-layer region. They are a mixture of wave-like Kelvin-Helmholtz structures (based on velocity gradients) and mushroom-like Rayleigh-Taylor vortices (governed by density gradients).

The applied numerical scheme is based on finite differences. Using compact finite differences tridiagonal sets of equations have to be solved employing the pipelined Thomas algorithm in order to compute various spatial derivatives. In contrast to the NEC-SX8/9 vector machines with few, powerful compute nodes the solution of this tridiagonal system turned out to be a major bottleneck on the massively parallel Cray-XE6 system when using a large number of MPI tasks. Hence, in order to avoid processor idling fully explicit and sub-domain compact finite differences are implemented. The solution of the tridiagonal set of equations can now be obtained separately for each sub-domain, or even becomes obsolete in case of the explicit finite differences. This results in a high parallel efficiency for a large number of MPI tasks and, thus, a significant decrease of compute time. The numerical results for the wall-parallel cooling gas injection showed a very good agreement between the three different finite-difference schemes, if all give a solution. The compact scheme is much more robust and can be used with a somewhat coarser grid. Thus, the sub-domain compact scheme is our method of choice.

Future investigations will account for main flow turbulence and 3-d cooling channels, and are aimed at identifying ineffective cooling setups. If the domain decomposition in streamwise and wall-normal direction of the present 2-d simulations is kept constant (2,500 MPI tasks), and if all 32 cores on each compute node are then used for OMP communication in spanwise direction, a total amount of 80,000 cores would have to be used. This corresponds to roughly 70 % of the whole system and, thus, a computation would be difficult due to long queuing times. Therefore, only four OMP threads will be used, which, according to Fig. 18, results in the best performance for the NS3D code. The figure illustrates a comparison of

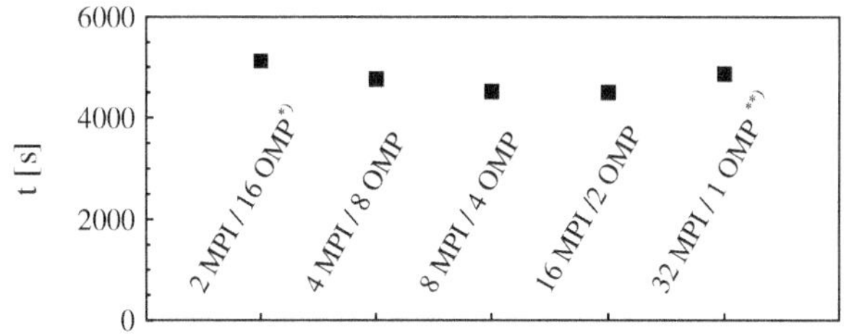

Fig. 18 MPI vs. OMP communication on one single compute node for a 3-d test problem with a total domain size of $320(x)\times100(y)\times128(z)$, 2,000 time steps. *Mixed MPI/OMP communication using $160(x) \times 100(y) \times 128(z)$ grid points per MPI task. **Pure MPI communication using $10(x) \times 100(y) \times 128(z)$ grid points per MPI task

OMP versus MPI communication, where a given three-dimensional test problem with $320(x) \times 100(y) \times 128(z)$ grid points is distributed on one compute node using pure MPI or mixed MPI/OMP communication. Note that the case with pure OMP communication could not be computed due to memory problems. The figure demonstrates that the best performance is obtained for a mixed MPI/OMP distribution with eight MPI tasks and four OMP threads, respectively, probably due to the NUMA-node architecture. However, using more OMP threads is just marginally slower indicating that the use of 16 or even 32 OMP threads is viable.

Due to geometrical requirements it is also planned to implement a domain decomposition in spanwise direction using MPI communication. Furthermore, the sub-domain compact approach is going to be applied to the numerical filtering procedure in streamwise and wall-normal direction allowing to reduce the spatial resolution, and eventually leading to less costly large-eddy simulations where appropriate. Here, also a tridiagonal set of equations has to be solved, since a compact filter of 10th-order is used [17].

Acknowledgements This work was funded by the German Research Foundation (Deutsche Forschungsgemeinschaft) in the framework of the Collaborative Research Center SFB/TRR 40: Fundamental technologies for the development of future space-transport-system components under high thermal and mechanical loads. Computational resources were kindly provided by the Federal High Performance Computing Center Stuttgart (HLRS) within project LAMTUR.

References

1. Aupoix, B., Mignosi, A., Viala, S., Bouvier, F., Gaillard, R.: Experimental and Numerical Study of Supersonic Film Cooling. AIAA J. **36/6**, 915–923 (1998) doi: 10.2514/2.495
2. Babucke, A., Linn, J., Kloker, M.J., Rist, U.: Direct Numerical Simulation of Shear Flow Phenomena on Parallel Vector Computers. In: M. Resch et al. (eds) High Performance Computing on Vector Systems, 229–247 (2003) doi: 10.1007/3-540-35074-8_16
3. Babucke, A.: Direct Numerical Simulation of Noise Generation Mechanisms in the Mixing Layer of a Jet. PhD Thesis, University of Stuttgart, Germany (2009)

4. Dahmen, W., Gotzen, T., Müller, S.: Numerical Simulation of Cooling Gas Injection Using Adaptive Multiscale Techniques. V European Conference on Computational Fluid Dynamics, Lisbon, Portugal (2010)
5. Gülhan, A., Braun, S.: An Experimental Study on the Efficiency of Transpiration Cooling in Laminar and Turbulent Hypersonic Flows. Exp. Fluids **50/3**, 509–525 (2011) doi: 10.1007/s00348-010-0945-6
6. Haidn, O.J.: Advanced Rocket Engines. In: Advances on Propulsion Technology for High-Speed Aircraft, 6-1–6-40 (1992)
7. Heufer, K.A., Olivier, H.: Experimental Study of Active Cooling in 8 Laminar Hypersonic Flows. In: Gülhan A. (eds) RESPACE - Key Technologies for Reusable Space Systems, Notes on Numerical Fluid Mechanics and Multidisciplinary Design, **98** 132–150 (2008) doi: 10.1007/978-3-540-77819-6_8
8. Heufer, K.A., Olivier, H.: Experimental and Numerical Study of Cooling Gas Injection in Laminar Supersonic Flow. AIAA J. **46/11**, 2741–2751 (2008) doi: 10.2514/1.34218
9. Juhany, K.A., Hunt, M.L.: Flowfield Measurement in Supersonic Film Cooling Including the Effect of Shock-Wave Interaction. AIAA J. **32/3**, 578–585 (1994) doi: 10.2514/3.12024
10. Kloker, M.: A Robust High-Resolution Split-Type Compact FD Scheme for Spatial Direct Numerical Simulation of Boundary-Layer Transition. Appl. Sci. Res. **59**, 353–377 (1998)
11. Konopka, M., Meinke, M., Schröder, W.: Large Eddy Simulation of Supersonic Film Cooling at Finite Pressure Gradients. In: W.E. Nagel et al. (eds) High Performance Computing in Science and Engineering '11, 353–369 (2011) doi: 10.1007/978-3-642-23869-7_26
12. Lele, S.K.: Compact Finite Difference Schmes with Spectral-like Resolution. J. Comput. Phys. **103**, 16–42 (1992)
13. Linn, J., Kloker, M.J.: Numerical Investigations of Film Cooling. In: Gülhan A. (eds) RESPACE - Key Technologies for Reusable Space Systems, Notes on Numerical Fluid Mechanics and Multidisciplinary Design, **98** 151–169 (2008) doi: 10.1007/978-3-540-77819-6_9
14. Linn, J., Kloker, M.J.: Effects of Wall-Temperature Conditions on Effusion Cooling in a Mach-2.67 Boundary Layer. AIAA J. **49/2**, 299–307 (2011) doi:10.2514/1.J050383
15. Linn, J.: Numerical Investigations of Film Cooling in Laminar Supersonic and Hypersonic Boundary-Layer Flows (Numerische Untersuchungen zur Filmkühlung in laminaren Über- und Hyperschallgrenzschichtströmungen). PhD Thesis, University of Stuttgart, Germany (2011)
16. Povitsky, A., Morris, P.: A Higher-Order Compact Method in Space and Time Based on Parallel Implementation of the Thomas Algorithm. J. Comput. Phys. **161**, 182–203 (2000) doi:10.1006/jcph.2000.6497
17. Visbal, M.R., Gaitonde, D.V.: On the use of High-Order Finite Difference Schemes on Curvilinear and Deforming Meshes. J. Comput. Phys. **181**, 155–185 (2002)
18. White, F.M.: Viscous Fluid Flow. McGraw-Hill (1991)
19. Winterfeldt, L., Laumert, B., Tano, R., James, P., Geneau, F., Blasi, R., Hagemann, G.: Redesign of the Vulcain 2 Nozzle Extension. 41st AIAA/ASME/SAE/ASEE Joint Propulsion Conference and Exhibit, AIAA Paper 2005–4536, Tucson, AZ, USA (2005)

Large Scale Numerics Uncovering New States of Matter

A. Moreno, J.M.P. Carmelo, and A. Muramatsu

Abstract While in condensed matter systems the constituents are well known, namely electrons, neutrons and protons, their interplay may give rise to unexpected states of matter. In this contribution we concentrate on strongly correlated electrons in one dimension driven out of equilibrium. This requires in principle, the solution of Schrödinger's equation dealing with a space of states, whose dimension increases exponentially with the number of electrons. Implementing an algorithm that requires only polynomially increasing computational resources, namely the time-dependent density matrix renormalization group (t-DMRG), we show that an electron injected into the system, fractionalizes in several portions, some of them carrying charge but no spin, and others carrying the spin and partial charge, in spite of the electron being an elementary particle in isolation. The characterization of such a fractionalization of charge and spin was made possible by the access to HPC plattforms with large memory processors.

A. Moreno (✉)
Institut für Theoretische Physik III, Universität Stuttgart, Pfaffenwaldring 57, 70550 Stuttgart, Germany
e-mail: moreno@itp3.uni-stuttgart.de

J.M.P. Carmelo
Center of Physics, University of Minho, Campus Gualtar, P-4710-057 Braga, Portugal

Institut für Theoretische Physik III, Universität Stuttgart, Pfaffenwaldring 57, D-70550 Stuttgart, Germany

Beijing Computational Science Research Center, Beijing, 100084, China
e-mail: carmelo@fisica.uminho.pt

A. Muramatsu
Institut für Theoretische Physik III, Universität Stuttgart, Pfaffenwaldring 57, 70550 Stuttgart, Germany

Beijing Computational Science Research Center, Beijing, 100084, China
e-mail: mu@theo3.physik.uni-stuttgart.de

M.M. Resch et al. (eds.), *Sustained Simulation Performance 2013*,
DOI 10.1007/978-3-319-01439-5_9,

1 Introduction

Condensed matter systems are composed of very well known particles, namely electrons, protons and neutrons. Furthermore, their physical properties are determined solely by the Coulomb interaction. In spite of this apparent simplicity, condensed matter systems display an enormous spectrum of physical phenomena, like superconductivity, superfludity, and magnetism. In recent years, new phenomena like supersolidity [1] and superconductivity in iron compounds [2], and new states of matter like non-abelian states [3], topological insulators [4], or spin liquids [5], together with electronic systems in one dimension [6] became the center of attention in the discussion of exotic forms of matter [7].

In this contribution we concentrate on one-dimensional (1D) quantum systems, where recent experimental advances allowed to access phenomena like spin-charge separation and charge fractionalization [6]. At low energies these systems are well described by the Luttinger Liquid (LL) theory [8] that predicts two independent excitations carrying either only charge (holons) or only spin (spinons) and propagating with different velocities, and hence, spin-charge separation. Experimental evidences of its existence have been observed in quasi-1D organic conductors [9], semiconductor quantum wires [10], and quantum chains on semiconductor surfaces [11]. The LL theory also predicts the fractionalization of injected charge into two chiral modes (left- and right-going) [12], a phenomenon recently confirmed experimentally [13]. Along the experimental advances also theoretical progress was recently achieved pertaining extensions beyond the LL limit by incorporating nonlinearity of the dispersion, leading to qualitative changes in the spectral function [14–16] and relaxation processes of 1D electronic systems [17].

Here we review work where we showed that fractionalization of charge and spin beyond the forms described by LL theory takes place when a spin-1/2 fermion is injected into a strongly correlated 1D system, namely the t-J model [18]. By studying the time evolution of the injected wavepacket at different wavevectors k, using time-dependent density matrix renormalization group (t-DMRG) [19, 20] different regimes were obtained. When k is close to the Fermi wavevector k_F, the known features from LL theory like spin-charge separation and fractionalization of charge into two chiral modes result. On increasing k, a further fractionalization of charge and spin appears, in forms that depend on the strength of the exchange interaction J or the density n. Their dynamics can be understood at the supersymmetric (SUSY) point $J = 2t$ in terms of charge and spin excitations of the Bethe-Ansatz solution [21, 22]. For the region of the phase diagram [23, 24], where the ground state corresponds to a repulsive LL, two qualitatively different regimes are identified: one regime with $v_s > v_c$ and another where $v_s < v_c$. Here $v_{c(s)}$ is the velocity of the excitations mainly carrying charge (spin). For $v_s > v_c$ and $k > k_F$ the spin excitation starts to carry a fraction of charge that increases with k while v_c corresponds to a wavepacket carrying only charge. For $v_s < v_c$ and $k > k_F$ the situation is reversed and the fastest charge excitation carries a fraction of spin that increases with k while the wavepacket with v_s carries almost no charge, i.e. in this case spin fractionalizes.

2 Model and Algorithms

The Hamiltonian of the 1D t-J model is as follows,

$$H = -t \sum_{i,\sigma} \left(\tilde{c}^{\dagger}_{i,\sigma} \tilde{c}_{i+1,\sigma} + \text{h.c.} \right) + J \sum_{i} \left(\boldsymbol{S}_i \cdot \boldsymbol{S}_{i+1} - \frac{1}{4} n_i n_{i+1} \right), \qquad (1)$$

where the operator $\tilde{c}^{\dagger}_{i,\sigma}$ ($\tilde{c}_{i,\sigma}$) creates (annihilates) a fermion with spin $\sigma = \uparrow, \downarrow$ on the site i. They are not canonical fermionic operators since they act on a restricted Hilbert space without double occupancy. $\boldsymbol{S}_i = \tilde{c}^{\dagger}_{i,\alpha} \boldsymbol{\sigma}_{\alpha\beta} \tilde{c}_{i,\beta}$ is the spin operator and $n_i = \tilde{c}^{\dagger}_{i,\sigma} \tilde{c}_{i,\sigma}$ is the density operator.

We study the time evolution of a wavepacket, corresponding to a fermion with spin up injected into the ground state, by means of t-DMRG [19, 20], that extends the DMRG method [25, 26] originally developed for the ground-state of quasi-one dimensional systems in equilibrium, to the solution of the time dependent Schrödinger's equation. A review of the original method and later advances can be found in review articles by U. Schollwöck [27, 28].

The state of a gaussian wavepacket $|\psi\rangle$ centered at x_0, with width Δ_x and average momentum k_0, is created by the operator $\psi^{\dagger}_{\uparrow}$ applied onto the ground state $|G\rangle$:

$$|\psi\rangle \equiv \psi^{\dagger}_{\uparrow} |G\rangle = \sum_i \varphi_i \tilde{c}^{\dagger}_{i\uparrow} |G\rangle, \qquad (2)$$

with

$$\varphi_i = A e^{-(x_i - x_0)^2 / 2\Delta_x} e^{i k_0 x_i}. \qquad (3)$$

A is fixed by normalization. The time evolved state $|\psi(\tau)\rangle$ by the Hamiltonian (1) determines the spin (s) and charge (c) density relative to the ground state as a function of time τ measured in units of $1/t$ ($\hbar = 1$),

$$\rho_\alpha(x_i, \tau) \equiv \langle \psi(\tau) | n_{i\alpha} | \psi(\tau) \rangle - \langle G | n_{i\alpha} | G \rangle, \qquad (4)$$

where $\alpha = s, c$, $n_{ic} = n_{i\uparrow} + n_{i\downarrow}$, and $n_{is} = n_{i\uparrow} - n_{i\downarrow}$. Most of the numerical results were carried out on systems with $L = 160$ lattice sites, using 600 DMRG vectors (this translates into errors of the order of 10^{-4} in the spin and charge density up to times of $50/t$) and $\Delta_x = 5$ lattice sites (which corresponds to a width $\Delta_k \sim 0.06\pi$ in momentum space).

3 Results

We discuss first the time evolution of a wavepacket at the SUSY point $J = 2t$, since here we will be able to identify the different portions in which the wavepacket splits on the basis of the Bethe-Ansatz solution.

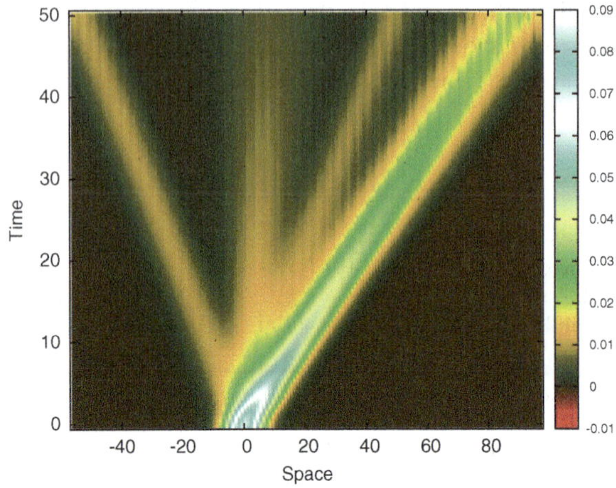

Fig. 1 Time evolution of $\rho_c(x_i, \tau)$ for a wavepacket initially at $x = 0$, with momentum $k = 0.7\pi$, at density $n = 0.6$, and $J = 2t$. Charge fractionalizes into four wavepackets, one to the *left*(P_1) and the rest(P_2, P_3, P_4) to the *right*. P_1 and P_3 have the same charge and speed but opposite velocities

Figure 1 shows the time evolution of $\rho_c(x_i, \tau)$ for a density of $n = 0.6$ ($n \equiv N/L$, with N the number of fermions). The momentum of the injected fermion is $k = 0.7\pi$, i.e. midway between $k_F = 0.3\pi$ and the zone boundary. The charge (i.e. ρ_c) splits into four fractions, one portion traveling to the left and the rest doing so to the right. A splitting into chiral modes is expected in the frame of LL theory [12], where for an injected right-going fermion, a splitting $Q_\alpha^{(\pm)} = (1 \pm K_\alpha)/2$ (where K_α is the so-called LL parameter and '+' ('−') corresponds to the right (left) propagating part) is predicted. The amount of charge (i.e. the integral of the wavepacket over its extension) corresponding to the portion denoted P_1 is $Q_c^{(-)} \sim 0.1$ which agrees well with the prediction of LL theory, since for the parameters in this case, $K_c \sim 0.8$ [24]. However, at long enough times, a further splitting of the right-going charge is observed (wavepackets P_i with $i = 2, 3, 4$), beyond the prediction of the LL theory.

Figure 2 displays both $\rho_c(x_i, \tau)$ (full line) and $\rho_s(x_i, \tau)$ (dashed line) for different values of the initial momentum of the injected wavepacket. The arrows indicate the direction of motion of each packet. As opposed to ρ_c, ρ_s does not split. In an SU(2) invariant system $K_s = 1$ [8], and hence $Q_s^{(-)} = 0$, i.e. no left propagating part is expected for the spin density. Moreover, part of the charge (P_4) is accompanying the spin, such that spin-charge separation does not appear to be complete. The amount of charge accompanying the spin increases as the momentum of the injected fermion approaches the zone boundary. These results make already evident that injecting a fermion at a finite distance from the Fermi energy leads to fractionalization of charge beyond the expectations from the LL theory.

In order to understand the new forms of fractionalization that go beyond the LL frame, we consider the excitations corresponding to one-particle addition processes, whose energies can be obtained from the Bethe-Ansatz solution [16, 29]. When adding an electron with momentum k, the single particle excitation energy is given by $\omega(k) = -\epsilon_c(q_c) - \epsilon_s(q_s)$, where $\epsilon_c(q_c)$ and $\epsilon_s(q_s)$ are the dispersion relations

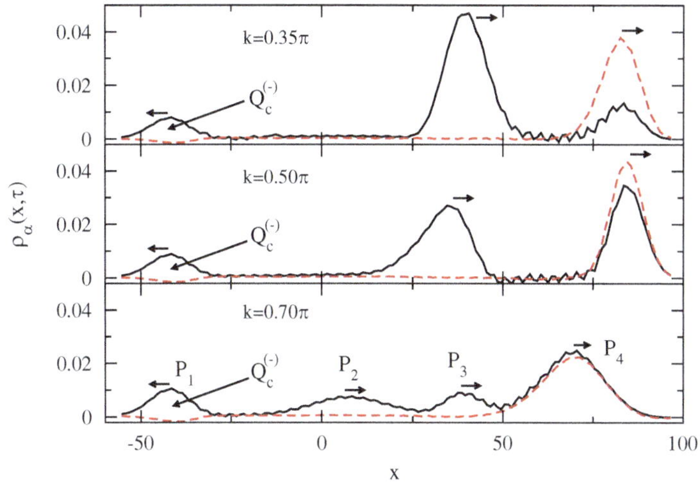

Fig. 2 Charge ($\rho_c(x_i, \tau)$, *full line*) and spin ($\rho_s(x_i, \tau)$, *dashed line*) densities for $J = 2t, n = 0.6$, at time $\tau = 40$, for different values of the momentum of the injected fermion

of the excitations for charge and spin, respectively, and the momenta are related to the momentum of the incoming particle as follows: $k = \pm 2k_F - q_c - q_s$, where $q_c \in [-q_{Fc}, q_{Fc}]$, and $q_s \in [-q_{Fs}, q_{Fs}]$, with $q_{Fc} = \pi - 2k_F$ and $q_{Fs} = \pi - k_F$ the pseudo-Fermi momenta for the excitations for charge and spin, respectively [22].

Figure 3 displays the velocities obtained from t-DMRG for the different wave-packets (symbols) compared to those obtained from Bethe-Ansatz (full lines), as a function of the momentum of the injected fermion. The velocity of each P_i is extracted by measuring the position of the maximum of the packet at the most convenient time, i.e., at that time where we can resolve P_i and the spreading of one packet does not destroy the other packets. The wavepackets P_1 (triangles) and P_3 (squares) have opposite directions, but the same speed and charges $Q_c^{(-)} \simeq Q_{c3}^{(+)}$, where the charges for the (+) branch are labelled by an index corresponding to the respective wavepackets. The velocity of the wavepacket P_4 (circles) agrees almost perfectly with the one corresponding to spin excitations. Its determination is best since it is the fastest wavepacket, such that it can be easily discerned from the rest. The velocity of the remaining wavepacket, P_2 (diamonds), is more difficult to assess, since it overlaps at the beginning with other ones. Nevertheless, its velocity closely follows the one of charge excitations. The wavepackets just described deliver a direct visualization of the excitations appearing in the Bethe-Ansatz solution, where only two different kind of particles are involved: the c and s pseudoparticles with their associated bands. The excitation associated with spin involves one hole in the c band with fixed momentum q_{Fc} and one hole in the s band with momentum q_s, where $q_s = \pm 2k_F - q_{Fc} - k$ [22]. In fact, the velocities of P_1 and P_3 correspond to the group velocity at both pseudo-Fermi momenta $\pm q_{Fc}$. Furthermore, as shown in Fig. 3, the velocity of those fractions is independent of the momentum of the injected

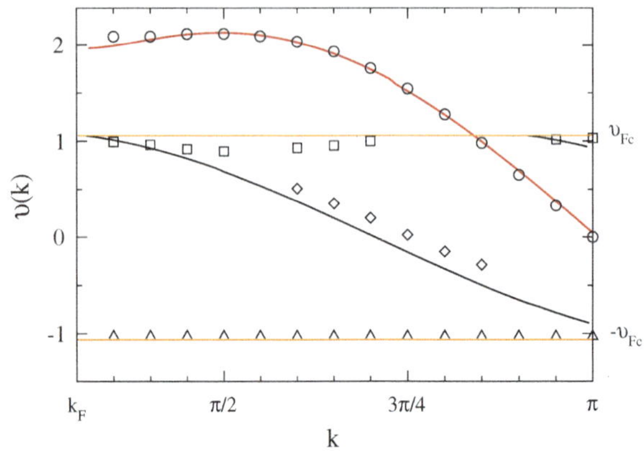

Fig. 3 *Full lines*: derivatives $v_\alpha(k) = \partial\epsilon_\alpha(k)/\partial k$ of the dispersions obtained by the Bethe-Ansatz solution. The symbols correspond to the velocities of the different wavepackets identified in Fig. 2: *triangles*(P_1), *diamonds*(P_2), *squares*(P_3) and *circles*(P_4). The *orange horizontal lines* stand for the Fermi velocity $v_{Fc} = \partial\epsilon_c(q_{Fc})/\partial q$ given by the Bethe-Ansatz solution

fermion, in agreement with the picture given by Bethe-Ansatz. The dispersion of the hole in the s-band gives rise to the velocity displayed by the red line in Fig. 3. Similarly, the c line (black line in Fig. 3) involves one hole in the s band with fixed momentum q_{Fs} and one hole in the c band with momentum q_c determined in terms of k by $q_c = \pm 2k_F - q_{Fs} - k$. Using the same argument as for the s line we can associate the P_2 packet (diamonds) with the c pseudoparticle. However, in this case we cannot observe wavepackets associated with spin and velocities corresponding to the group velocity at the pseudo-Fermi momenta $\pm q_{Fs}$. We understand this as due to the fact that $K_s = 1$, by analogy to what we observe in the $K_c = 1$ case. On the SUSY point this case is reached in the limit of vanishing density, where the system can be described by a Fermi gas. Hence, fractionalization is absent in this limit.

Next we depart from the SUSY point and examine how fractionalization takes place in the region of the phase diagram where the ground state corresponds to a LL with $K_c < 1$. As shown in Fig. 4, essentially the same features are observed as at the SUSY point both for $J > 2t$ and $J < 2t$. In all the cases shown in Fig. 4, where the velocity of spin excitations (v_s) remains higher than that of charge excitations (v_c) in most parts of the Brillouin zone, spin does not fractionalize, as opposed to charge, so that the interpretation derived from Bethe-Ansatz remains valid over an extended region of the phase diagram: charge splits into four portions of which one travels with the spin wavepacket, and two have the same speed but opposite group velocity which does not depend on the momentum of the injected fermion. It is tempting to assign those excitations to states at a pseudo-Fermi surface for charge excitations.

In summary, we have shown through the time evolution of an injected spin-full fermion onto the t-J model, that charge and spin fractionalization occurs beyond the predictions of the Luttinger liquid theory. A comparison with results from Bethe-Ansatz allowed to identify charge and spin excitations that split into

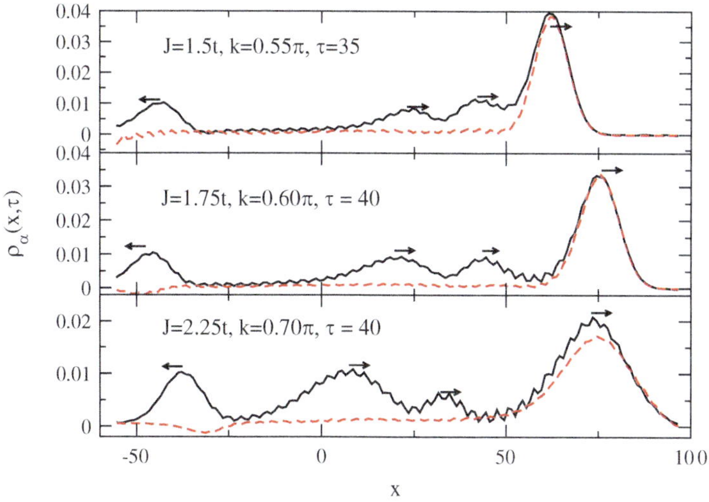

Fig. 4 Fractionalized wavepackets for different values of J/t away from the SUSY point, at a density $n = 0.6$. As in the SUSY case, charge fractionalizes into four pieces, while spin does not, and carries an appreciable amount of charge

components at high and low energies. The components at high energy reveal the dispersion ϵ_c and ϵ_s of charge and spin excitations, respectively. The components at low energy have a velocity that does not depend on the momentum of the injected fermion and are very well described by states at the pseudo-Fermi momenta of the charge excitation. This picture can be extended to a wide region in the phase diagram of the t-J model as long as the ground state corresponds to $K_c < 1$, and $v_s > v_c$. In this region fractionalization is observed only in the charge channel. However, for $v_c > v_s$, a region that develops for J/t below $\sim$1.5, the spin density shows fractionalization [18]. All over, the fastest excitation is accompanied by the complementary one, such that spin-charge separation is for them only partial. The other fractions present an almost complete spin-charge separation.

Acknowledgements A. M. and A. M. acknowledge support by the DFG through SFB/TRR 21. A. M. and J. M. P. C. thank the hospitality and support of the Beijing Computational Science Research Center, where part of the work was done. J. M. P. C. thanks the hospitality of the Institut für Theoretische Physik III, Universität Stuttgart, and the financial support by the Portuguese FCT both in the framework of the Strategic Project PEST-C/FIS/UI607/2011 and under SFRH/BSAB/1177/2011, the German transregional collaborative research center SFB/TRR21, and Max Planck Institute for Solid State Research. A. M. thanks the KITP, Santa Barbara, for hospitality. This research was supported in part by the National Science Foundation under Grant No. NSF PHY11-25915. We are very grateful to HLRS (Stuttgart) and NIC (Jülich) for providing the necessary supercomputer resources.

References

1. Balibar, S.: The enigma of supersolidity. Nature **464** pp. 176–182, (2010)
2. Mazin, I.: Superconductivity gets an iron boost. Nature **464** pp. 183–186, (2010)
3. Stern, A.: Non-Abelian states of matter. Nature **464** pp. 187–193, (2010)
4. Moore, C.: The birth of topological insulators. Nature **464** pp. 194–198, (2010)
5. Balents, L.: Spin liquids in frustrated magnets. Nature **464** pp. 199–208, (2010)
6. Deshpande, V. V., Bockrath, M., Glazman, L. I., and Yacoby, A.: Electron liquids and solids in one dimension. Nature **464** pp. 209–216, (2010)
7. Csontos, D.: Exotic matter. Nature **464** pp. 175, (2010)
8. T. Giamarchi, *Quantum Physics in One Dimension* (Clarendon Press, Oxford, 2004).
9. T. Lorenz, M. Hofmann, M. Gruninger, A. Freimuth, G. S. Uhrig, M. Dumm, and M. Dressel, Nature **418**, 614 (2002).
10. O. M. Auslaender, H. Steinberg, A. Yacoby, Y. Tserkovnyak, B. I. Halperin, K. W. Baldwin, L. N. Pfeiffer, and K. W. West, Science **308**, 88 (2005).
11. C. Blumenstein, J. Schäfer, S. Mietke, A. Dollinger, M. Lochner, X. Y. Cui, L. Patthey, R. Matzdorf, and R. Claessen, Nature Phys. **7**, 776 (2011).
12. K.-V. Pham, M. Gabay, and P. Lederer, Phys. Rev. B **61**, 16397 (2000).
13. H. Steinberg, G. Barak, A. Yacoby, L. N. Pfeiffer, K. W. West, B. I. Halperin, and K. L. Hur, Nature Phys. **4**, 116 (2008).
14. A. Imambekov and L. I. Glazman, Science **323**, 228 (2009).
15. A. Shashi, L. I. Glazman, J.-S. Caux, and A. Imambekov, Phys. Rev. B **84**, 045408 (2011).
16. J. M. P. Carmelo, K. Penc, and D. Bozi, Nucl. Phys. B **725**, 421 (2005); **737**, 351 (2006).
17. G. Barak, H. Steinberg, L. N. Pfeiffer, K. W. West, L. Glazman, F. von Oppen, and A. Yacoby, Nature Phys. **6**, 489 (2010).
18. A. Moreno, A. Muramatsu, and J. M. P. Carmelo, Phys. Rev. B **87**, 075101 (2013).
19. S. R. White and A. E. Feiguin, Phys. Rev. Lett **93**, 076401 (2004).
20. A. J. Daley, C. Kollath, U. Schollwöck, and G. Vidal, J. Stat. Mech.: Theor. Exp. P04005 (2004).
21. P. A. Bares, G. Blatter, and M. Ogata, Phys. Rev. B **44**, 130 (1991).
22. P. A. Bares, J. M. P. Carmelo, J. Ferrer, and P. Horsch, Phys. Rev. B **46**, 14624 (1992).
23. M. Ogata, M. Luchini, S. Sorella, and F. Assaad, Phys. Rev. Lett **66**, 2388 (1991).
24. A. Moreno, A. Muramatsu, and S. R. Manmana, Phys. Rev. B **83**, 205113 (2011).
25. S. R. White, Phys. Rev. Lett **69**, 2863 (1992).
26. S. R. White, Phys. Rev. B **48**, 10345 (1993).
27. U. Schollwöck, Rev. Mod. Phys. **77**, 259 (2005).
28. U. Schollwöck, Ann. Phys. **326**, 96 (2011).
29. J. M. P. Carmelo, L. M. Martelo, and K. Penc, Nucl. Phys. B **737**, 237 (2006).

Towards Simulation of Electrodialytic Sea Water Desalination

Kannan Masilamani, Jens Zudrop, and Sabine Roller

Abstract Electrodialyis is a process engineering technique in which electrochemical potentials are used to separate different substances. It has been used in a couple of engineering application areas for many years already. Recently, sea water desalination by an electrodialytic process became of interest, as the concept offers a very efficient way for the generation of drink water from sea water. Although this way of drink water generation is known to be energetic efficient, the basic mechanisms in a desalination plant are not well known. In this paper we describe our approach to integrated, multiphysics simulation of such a process. We focus on the hydrodynamic part of a desalination stack and present a scalable simulation environment for mass transfer and hydrodynamics by means of a multicomponent Lattice Boltzmann method. Performance measurements on state of the art for realistic simulation setups on supercomputers are presented.

1 Introduction

Scarcity of drink water is a continuously growing problem in many country sides. For this reason desalination of sea water is attracting a lot of notice in the last years. A number of different process engineering techniques have been developed

K. Masilamani (✉)
Siemens AG, Corporate Technology, CT RTC ENC ENT-DE, Günther-Scharowsky-Str. 1, 91058 Erlangen, Germany

Simulationstechnik und wissenschaftliches Rechnen, University of Siegen, Hölderlinstr.3, 57076 Siegen, Germany
e-mail: kannan.masilamani@uni-siegen.de

J. Zudrop · S. Roller
Simulationstechnik und wissenschaftliches Rechnen, University of Siegen, Hölderlinstr.3, 57076 Siegen, Germany
e-mail: jens.zudrop@uni-siegen.de; sabine.roller@uni-siegen.de

M.M. Resch et al. (eds.), *Sustained Simulation Performance 2013*,
DOI 10.1007/978-3-319-01439-5_10,

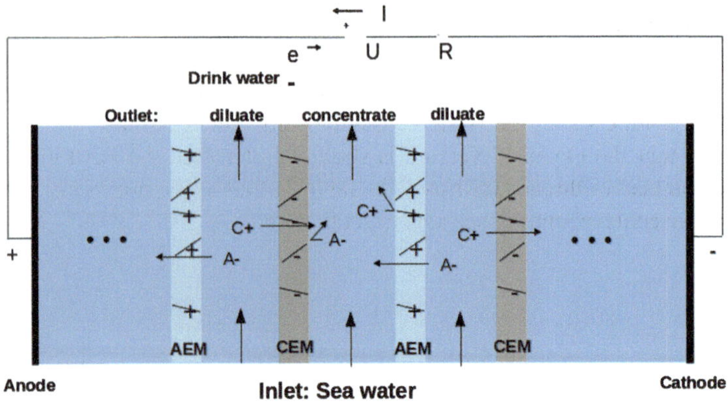

Fig. 1 Simplified structure of a desalination stack in two dimensions. A periodic arrangement of anion-exchange-membrane (AEM), diluate channel, cathion-exchange-membrane (CEM) and concentrate channel is embedded between anode and cathode. The sea water is fed in the diluate channel. After removing the ions, drink water can be extracted at the end of the diluate channel

to achieve the production of drink water from sea water. However, in addition to the drink water production itself economical aspects are of great importance, especially as prices for fossile energies are also continuously growing. From that point of view, process enegineering techniques with the highest energetic efficiency are a desired goal.

In the last years electrodialysis became a potential candidate for an efficient sea water desalination process. A conceptual structure of a electrodialysis stack is shown in Fig. 1. It consists of a periodic arrangement of anion-exchange-membranes (AEM), diluate channel, cathion-exchange-membranes (CEM) and concentrate channels. The stack is surronded by anode and cathode. AEM and CEM are semi-permeable membranes which allow only anions/cathions to pass through the membrane. By the electric potential (due to the voltage drop between anode and cathode) the ions of sea water, fed in the diluate channel, start to separate and move through the semi-permeable membranes, AEM and CEM respectively. After the separation of ions from sea water, desalinated water can be extracted at the end of diluate channel. To produce water that is drinkable by humans, sea water has to pass the desalination stack several times before the amount of salts is reduced to a acceptable level.

From the technical point of view, some additional complications arise. To stabilize the fluid channels (i.e. diluate and concentrate channel) mechanically, additional spacer structures have to be added to the channels. This strucutre is an interwoven mesh which allows the water to pass through the channel, while separating the membranes next to the fluid channel. Figure 2 shows a potential structure of such a spacer.

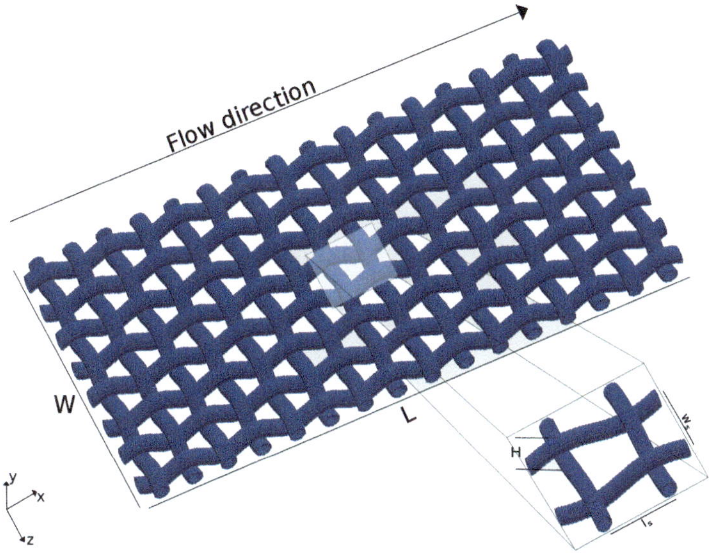

Fig. 2 Structure of a spacer utilized to stabilize the fluid channels of the spacer stack mechanically. The structure of this spacer has a significant impact on the total energy consumption of the stack

The energy consumption of the system in mainly influenced by two aspects:

- Energy consumption of pumps to press the sea water through the desalination stack
- Electric energy consumption to sustain the electric potential across the desalination stack

The first point is mainly influenced by the structure of the spacer geometry, as the structure (and the inflow parameters like Reynolds number) determine the pressure loss along a fluid channel and therefore the necessary work of the pumps. Furthermore, the structure of the spacer influences the mixing of sea water in the channel and may increase or decrease the ion separation process in the channel.

In this paper we focus on efficient simulations of the mass transport and hydrodynamics in the spacer filled fluid channel by means of a Lattice Boltzmann method. The paper is organized as follows: In Sect. 2 we describe the mathematical model together with the Lattice Boltzmann model we used for the numerical simulations. In Sect. 3 we describe our Octree based implementation. Section 4 shows performance results on modern, state of the art HPC systems for realistic simulation setups. We conclude in Sect. 5 and give an outlook to future work in this research area.

2 Liquid Mixture Modelling

The physical processes in the spacer filled fluid channels mainly two phenomena:

- Mass transport, with external driving forces due to the electric potential
- Hydrodynamics at low Mach number (i.e. in the incompressible regime)

Our partial differential equation based model is based on a momentum equation for the mixture of water and ions (i.e. u represents the velocity field of the mixture)

$$\partial_t u + (u \cdot \nabla) u = \nabla p + \nu \Delta u + F \tag{1}$$

and an additional mass transport equation for each substance (where n_i denotes the number density of substance i and w the molar averaged velocity of the mixture)

$$\partial_t n_i + \nabla \cdot n_i w = -\nabla J_i \tag{2}$$

modelling the behavior of each substance with respect to the mixture reference frame. The model is closed by the so-called Maxwell-Stefan model for the diffusive fluxes J_i. The Maxwell-Stefan diffusion

$$\nabla \chi_i + F_i = \sum_{l \neq i} \frac{1}{D_{i,l} n} (\chi_i J_l - \chi_l J_i) \tag{3}$$

determines the interaction of the different substances with each other and takes into account chemical and electrical potentials. We used the following variables in the Maxwell-Stefan formulation: The molar concentration χ_i for substance i, $D_{i,l}$ the diffusivities between species i and l and n the total molar concentration of the mixture. External driving forces for the mixtures or each substance separately are taken into account by F or F_i respectively.

The upper equations are relatively complex due to its nonlinear behavior and the fact that the number of partial differential equations is proportional to the number of species in the system. Furthermore, the spacer geometry is complex. Due to this reasons Lattice Boltzmann methods are a potential candidate for such simulation tasks. The next chapter briefly describes the Lattice Boltzmann model which we used for the simulations.

2.1 *Lattice Boltzmann Mixture Modelling*

Lattice Boltzmann Methods (LBM) are gas kinetic methods used to simulate the behavior of incompressible fluids. The methods can be seen as a particular discretization of the Boltzmann equation by integration along the characteristics, restriction to a lattice with integer positions (i.e. $x, x+1, \cdots$) and restriction to a finite number of velocities in phase space (e.g. c_k with $k = 0, \cdots, 8$ for the D2Q9 model). The usual LBM formulation (for a single substance) is given by:

$$f_k(x + c_k, t + 1) = \underbrace{f_k(x,t) + \underbrace{\lambda \left(f_k^{eq}(x,t) - f_k(x,t) \right)}_{C}}_{S}$$

f^{eq} represents the equilibrium distribution. Due to the structure of the LBM it can be subdivided into a collision and streaming step, i.e. C and S. The moments of f are related to the macroscopic fluid variables, e.g.

$$u = \sum_k c_k f_k.$$

The incompressible Navier-Stokes equation is recovered in the asymptotic limit of vanishing grid size. The so-called relaxation parameter λ is directly related to the fluid's kinematic viscosity ν.

In case of a mixture of multiple substances the Lattice Boltzmann method recovering Eqs. 1–3 becomes more complex. The model is based on [2, 3] and considers a probability density function f^i for each substance i. The interaction of the different substances is taken into account by a modified eqilibrium distribitution function $f^{i,eq}$. On the continuous level the model for substance i reads

$$\partial_t f_k^i(x,t) + c_k \cdot \nabla_x f_k^i(x,t) = \lambda \left(f^{i,eq} - f_k^i \right) \tag{4}$$

The equilibrium function is given by

$$f_k^{i,eq} = \omega_k \rho_k \left(s_i^k + \frac{1}{c_s^2} \left(c_k \cdot u^{i,*} \right) + \frac{1}{2c_s^4} \left(c_k \cdot u^{i,*} \right)^2 - \frac{1}{2c_s^2} \left(u^{i,*} \cdot u^{i,*} \right) \right) \tag{5}$$

and the modified velocity $u^{i,*}$ of species i by

$$u^{i,*} = u^i + \sum_l \frac{B_{i,l}}{B} \phi_i \chi_l \left(u_l - u_i \right) \tag{6}$$

For further details and tuning of the constants in the upper equilibrium and the velocity $u^{i,*}$ we refer again to [2]. This kinetic scheme is able to resolve mass transport and hydrodynamics at the same time in a single scheme, while keeping the simple structure of a stream and collide algorithm.

A discretization of the upper continuous equation can be achieved by integration along the characteristics. At the end an explicit algorithm (similar to the original stream and collide mechanism) can be obtained. For further details we refer to [3]. However, when compared to the "usual" Lattice Boltzmann model there is one major difference. In each collision step a cell local linear equation system has to be

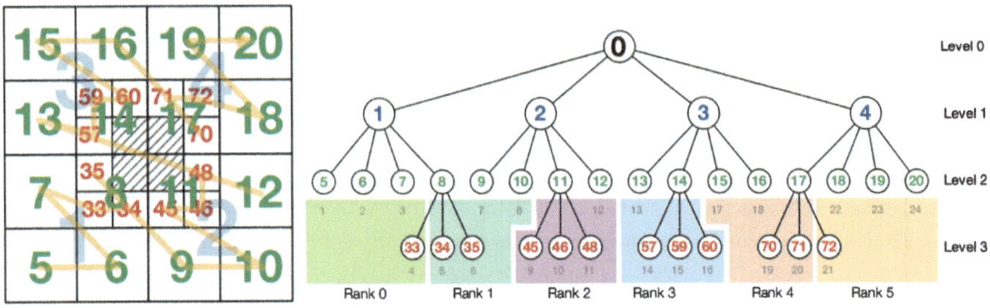

Fig. 3 Quadtree with breadth-first numbering scheme for the elements (*left*) and its tree representation with space-filling curve linearization of the leaf elements (*right*)

solved (to obtain the macroscopic velocities). It should be emphasized that the linear equation system is well-posed, its size corresponds to the number of substances and can be solved within each element alone. Therefore, this step has no negative impact on the scalablity of the method.

The next section describes the Octree based computational framework on which our implementation of the upper method is based.

3 Octree Based Simulations

As pointed out in the previous section, Lattice Boltzmann methods rely on a cartesian voxelization of the computational domain. Our implementation relies on an Octree representation of the mesh [1]. We equip the hierarichal datastructure with a breadth-first numbering scheme for the elements and a linearization of the elements by a space filling curve. A two-dimensional quadtree with the breadth-first numbering is shown in Fig. 3. Please notice, that we store only leaf elements of the tree (which represent the actual fluid elements of the mesh).

The domain decomposition relies on an equal distribution of elements to the processes by cutting the linearized list of elements into parts of equal size. It worth emphasizing that our approach is completely parallel in the following sense: We avoid $\mathcal{O}(p)$ algorithms (where p denotes the number of processes), instead we rely on fast alogrithms leading to $\mathcal{O}(\log(p))$. Furthermore, a specific processes does not require knowledge about all other processes, which would lead to $\mathcal{O}(p)$ memory consumption. In our current implementation we store the first and last element IDs (integer) for each rank and all ranks, however this can be easily overcome be making use of a distributed datastructure, once the memory overhead becomes to large.

A large problem of modern parallel software is often the IO. Most often it is not scalable enough and might hinder to run very large simulations. In our framework we achieve a fully parallel IO by using prefix summations with complexity $\mathcal{O}(\log(p))$.

The features of the Octree based mesh are encapsulated in a library (including the construction of computing stencils, neighbor element search and build up of communication patterns), such that the Lattice Boltzmann solver is focussing on implementations of method specific routines. The performance of our multicomponent Lattice Boltzmann solver is presented in the next section.

4 Performance and Applications

This section describes the performance of Musubi with complex spacer geometry on the Cray XE6 system Hermit at HLRS. The Hermit system provides of 3,552 computing nodes with AMD Interlagos on two sockets where each socket has 16 cores resulting in 32 cores per node. For our performance analysis, up to 1,024 computing nodes or 32,768 cores are used. Only MPI parallelism is considered here. Both, intranode and internode performance are measured i.e. performance within a single node (up to 32 cores with as many MPI processes) and between multiple nodes. For scaling analysis, the problem size is increased from a single spacer element length of 0.2 cm (Zoomed part in Fig. 2) with 66,000 elements to the laboratory scale spacer length of 20 cm with 66 million elements. In the width of this channel slice, a periodic boundary is assumed.

In Lattice-Boltzmann codes the measurement of the lattice updates per second is commonly used to compare the performance. This number of lattice updates per second (LUPs) will be used in the following presentation, as it provides a direct estimation on how long a given simulation, that requires a certain number of lattice nodes and time step updates will take. We represent the behavior of the code in terms of performance per execution unit, that is per node or per core, to get a clearer impression of the performance independent of the number of used execution units. An ideal parallel execution is expected to just replicate the serial behavior on each execution unit. However, the execution performance is influenced by cache usage, non-computational implementation overheads, vector lengths, communication times and so on.

Our solver framework ensures almost perfect computational load balancing with the help of space-filling curve i.e. there is at most single element difference between the partitions. However, due to the irregular domain and the large number of walls in spacer filled channel, the communication surface between different processes might vary drastically, resulting in a large imbalance of communication costs. The effect of this load imbalance can be noticed in the performance map shown in Fig. 4b with a relatively high performance drop for smaller domains per node.

First, the performance runs of the single-fluid and multi-species LBM are performed using a single node with 32 cores, depicted in Fig. 4a. In both models a D3Q19 layout with a BGK-like collision operator is used. Streaming and collision steps are solved at each time step for both models. However, the multi-species LBM model used in this work requires the solution of an additional cell local linear equation system, increasing the required number of floating point operations

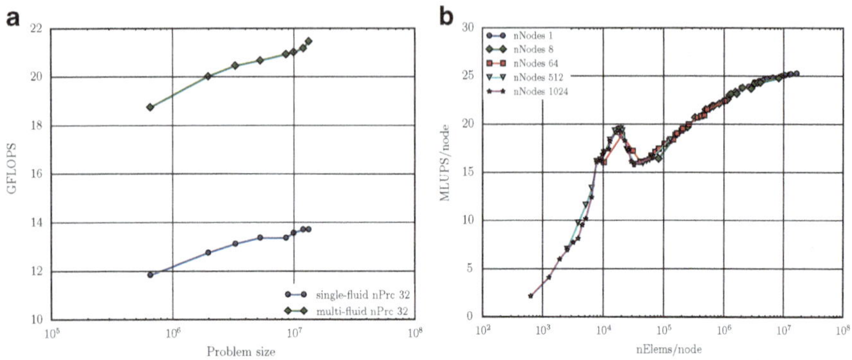

Fig. 4 Full node performance of single fluid and multi-species LBM model with spacer structureIntranode (**a**) and internode (**b**) perfromance map with spacer structure

per lattice update. The single fluid solver requires only about 150 floating point operations, while a simulation with 3 species requires with 850 operations more than just a factor of 3 more operations.

In our LBM solver, instead of communicating all probability density functions on the communication surface between the processors, only the required links of probability density functions are communicated. This reduces the MPI buffer size and bandwidth driven communication times. In Fig. 4a, We cannot see the cache region since smallest problem size i.e. single spacer element with 66,000 elements is above cache limit. The Hermit system has a theoretical peak performance of 294.4 GFLOPS per node. We achieved a sustained performance of roughly 4.2 % on a full node with single-fluid LBM and 7.2 % with multi-species LBM.

The internode performance map is shown in Fig. 4b with problem size per node in horizontal axis and performance per node in vertical axis for various total number of nodes. The performance map combines both, weak scaling and strong scaling. Weak scaling can be measured by the vertical comparison of points between different lines for the node count i.e. fixing the number of elements per node. The closer the points are located to each other the better the weak scaling. Strong scaling on the other hand is not as easily seen in the performance map, but can be derived by moving to the left when increasing the number of nodes. This reduces the number of elements per process with increasing node counts, as required by strong scaling with a fixed overall problem size.

The steep slope after the cache effect in both models Fig. 4b is supposedly due to load imbalances in the communication times, caused by the walls of the spacer geometry. This can be explained with the help periodic testcase performance Fig. 5, where the performance flattens out after the cache effect with a small slope. The main difference between the two simulations is the exactly balanced communication effort in the one case and a less than optimal balancing for the complex spacer geometry. This can be resolved using a dynamic load balancing algorithm to

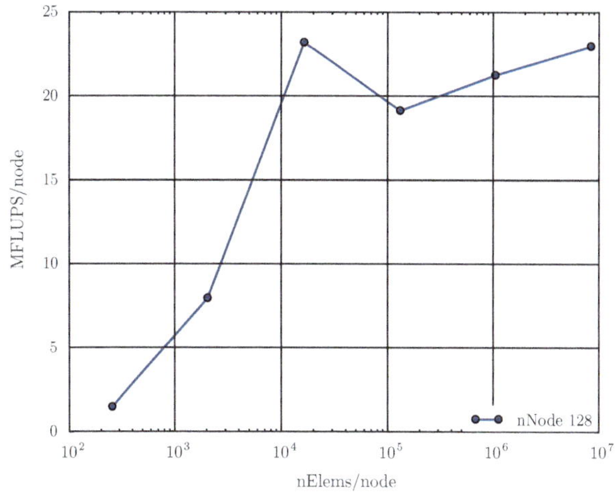

Fig. 5 Performance of the multi-species LBM implementation on the Hermit system on 128 nodes with periodic cubic simulation domain

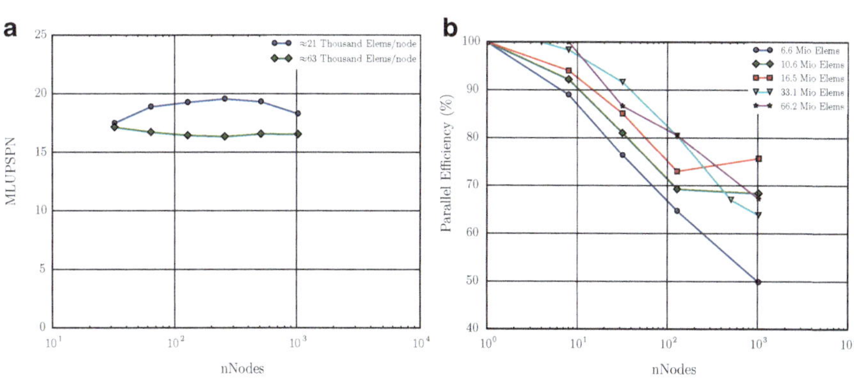

Fig. 6 **(a)** Weak scaling for two different number of elements per node with spacer structure for different number of nNodes. Ideal weak scaling is a straight line. **(b)** Strong scaling parallel efficiency for different problem sizes with spacer structure

distribute the simulation domain on each processor at runtime according to the actual load.

From Fig. 4b, it can be seen that weak scaling works fine on different process counts for all problems, that fit into memory down to the cache-sized problems, where the communication gets dominant and the performance per node drops down with a steeper slope.

An explicit plot for weak scaling for two different problem size per node is shown in Fig. 6a. Here, 21,000 elements per node fit into the cache and 63,000 elements is above the cache limit. Figure shows that weak scaling is almost perfect with only a small drop in the performance for larger counts of compute nodes.

A dedicated graph for strong scaling is shown in Fig. 6b with number of nodes on the horizontal axis and parallel efficiency (%) in the vertical axis. Here, testcases with problem sizes of roughly 6.6, 10.6, 16.5 and 66.2 million elements are used. Here, with a problem size of 6.6 million elements, the performance drops from 1 node to 1,024 nodes because the communication dominates the computation. The peak in the plots defines the problems which fits in the cache. The performance efficiency of multi-species LBM for large problem sizes with 1,024 nodes is less than for smaller problem sizes because it did not fit into the cache and is at the local performance minimum immediately after leaving the cache instead. Hence, to fit the problem with full spacer length into cache, compute nodes of roughly 2 or 3 times more than 1,024 nodes are required. Otherwise, the problem size per node should be increased to gain higher performance levels in the region where the influence of the communication is negligible.

5 Conclusion and Outlook

In this paper, we presented the multi-species lattice Boltzmann method for liquid mixture modeling to simulate the transport of ions and mixture in the spacer filled flow channel in electro-dialytic process. We implemented this model in our lattice Boltzmann solver framework and presented the performance of our implementation for mixture with three species flow with complex spacer geometry. The scaling analysis shows that we can scale up to thousand of cores and can simulate full length laboratory scale spacer. In the future, we will investigate on the special boundaries for mixture diffusive fluxes to treat membranes. Regarding the performance, the dynamic load balancing to reduce communication time will be investigated

Acknowledgements This work was funded by the German Federal Ministry of Education and Research (Bundesministerium für Bildung und Forschung, BMBF) in the framework of the HPC software initiative in the project HISEEM.

References

1. S. Roller, J. Bernsdorf, H. Klimach, M. Hasert, D. Harlacher, M. Cakircali, S. Zimny, K. Masilamani, L. Didinger, J. Zudrop: An Adaptable Simulation Framework Based on a Linearized Octree. In: M. Resch, X. Wang, W. Bez, E. Focht, H. Kobayashi, S. Roller (eds.) High Performance Computing on Vector Systems 2011, Springer Berlin Heidelberg 2012
2. J. Zudrop, S. Roller., P. Asinari: A Lattice Boltzmann scheme for liquid mixtures - Part I: Model and analysis. submitted to Phys. Rev. E (2013)
3. J. Zudrop, S. Roller., P. Asinari: A Lattice Boltzmann scheme for liquid mixtures - Part II: Discretization and numerics. submitted to Phys. Rev. E (2013)

A Regional Climate Model Simulation for EURO-CORDEX with the WRF Model

Kirsten Warrach-Sagi, Thomas Schwitalla, Hans-Stefan Bauer, and Volker-Wulfmeyer

Abstract In order to provide high-resolution ensembles and comparisons of regional climate simulations, the World Climate Research Program (WCRP) initiated the COordinated Regional climate Downscaling Experiment (CORDEX). CORDEX is performed in preparation of the fifth assessment report of the Intergovernmental Panel on Climate Change (IPCC AR5) (Giorgi et al., WMO Bull 58:175–183, 2009). Verification runs for CORDEX are performed for most continents for a 20-year period (1989–2009) driven by ERA-interim data from the European Centre for Medium Range Weather Forecast (ECMWF). For Europe (EURO-CORDEX, http://www.euro-cordex.net) an ensemble of regional climate model simulations from 1989 to 2008 on 0.11°, 0.22° and 0.44° has been completed in May 2012. The University of Hohenheim contributed to EURO-CORDEX with a simulation with Weather Research and Forecast (WRF) model on the CRAY XE6 of the High Performance Computing Center Stuttgart (HLRS) of the University of Stuttgart. The model consists of a spatial grid of 424*412*54 grid cells and is run with a timestep of 60 s on 1,280 processors. Three-hourly output of the atmospheric and terrestrial variables is written to daily netcdf-files each of the size of 8.5 GB. The simulations' set up is described and a comparison of the results to an observational precipitation data set for Germany is shown.

1 Introduction

The application of numerical modeling for climate projections is an important task in scientific research since they are the most promising means to gain insight in possible future climate changes. The quality of the prepared global projections has been

K. Warrach-Sagi (✉) · T. Schwitalla · H.-S. Bauer · Volker-Wulfmeyer
Institut für Physik und Meteorologie, Universität Hohenheim, Garbenstr. 30, 70599 Stuttgart, Germany
e-mail: Kirsten.Warrach-Sagi@uni-hohenheim.de; thomas.schwitalla@uni-hohenheim.de; hans-stefan.bauer@uni-hohenheim.de; volker.wulfmeyer@uni-hohenheim.de

M.M. Resch et al. (eds.), *Sustained Simulation Performance 2013*,
DOI 10.1007/978-3-319-01439-5_11,

continuously improved in recent years, enabled by more powerful supercomputers as well as advanced numerical and physical schemes (e.g. [20, 22, 23]). During the last two decades, various regional climate models (RCM) have been developed and applied for simulating the present and future climate of Europe. First of all, the performance of the RCMs to successfully reproduce the observed regional climate characteristics within the last decades was extensively assessed. Within the EU projects ENSEMBLES and PRUDENCE, ensemble simulations of RCMs forced with ERA-40 reanalysis data were executed and analyzed with a grid resolution of the order of 50 km [7, 8]. It was found that these models were able to reproduce the pattern of temperature distributions reasonably well but a large variability was found with respect to the simulation of precipitation. The performance of the RCMs was strongly dependent on the quality of the boundary forcing, namely if precipitation was due to large-scale synoptic events. Additionally, summertime precipitation was subject of significant systematic errors as models with coarse grid resolution have difficulties to simulate convective events. This resulted in deficiencies with respect to simulations of the spatial distribution and the diurnal cycle of precipitation. Correspondingly, the 50-km resolution RCMs were hardly capable of simulating the statistics of extreme events such as flash floods. The results of ENSEMBLES and PRUDENCE are in accordance with a variety of studies of RCMs at the order of 25–50 km (e.g., [1, 10, 18, 19]). Due to these reasons, RCMs with higher grid resolution of 10–20 km were developed and extensively verified, e.g. in southwest Germany [13, 14]. While still inaccuracies of the coarse forcing data were transferred to the results, these simulations indicated a gain from high resolution due to better resolution of orographic effects. These include an improved simulation of the spatial distribution of precipitation and wet day frequency and extreme values of precipitation. However, three major systematic errors remained: the windward-lee effect [13], phase errors in the diurnal cycle of precipitation [3], and precipitation return values, especially on longer return periods [14]. In order to provide high-resolution ensembles and comparisons of regional climate simulations, the World Climate Research Program (WCRP) initiated the COordinated Regional climate Downscaling EXperiment (CORDEX) has been initiated. CORDEX is performed in preparation of the fifth assessment report of the Intergovernmental Panel on Climate Change (IPCC AR5) [15]. Within WRFCLIM_project of the University of Hohenheim at HLRS a verification run for Europe was performed with the Weather Research and Forecasting (WRF) model [22] for a 20-year period (1989–2009) on the NEC Nehalem Cluster at a grid resolution of approx. 12 km with CORDEX simulation requirements. Precipitation and soil moisture results of this simulation were analysed [16, 26] and due to the results in the beginning of 2012 the simulation was repeated with an updated version of WRF on the CRAY X6 at HLRS. This simulation was completed in April 2012 and is currently under evaluation e.g. within the EURO-CORDEX (www.euro-cordex.net) ensemble of regional climate models (e.g. [25]). In the future regional climate projection runs (1950–2100) will be forced with the latest global climate model runs (CMIP5 data) in the frame of CORDEX, which currently become available to the regional climate modeling community.

Within this context the objectives are to

- Provide high resolution (12 km) climatological data to scientific community
- Support of quality assessment and interpretation of the regional climate projections for Europe through the contribution to the climate projection ensemble for Europe (EURO-CORDEX) for the next IPCC (International Panel of Climate Change) report.

Generally, three forcing conditions for the formation of precipitation in mid-latitude terrains with complex orography and land-surface heterogeneity can be distinguished [28]: (1) strongly-forced conditions, e.g., due to the presence of a surface front; (2) weakly-forced conditions, no surface front but upper-level instabilities, e.g., due to the presence of an upper-level trough and advection of potential vorticity; (3) air mass convection: no strongly-forced or weakly forced conditions, likelihood of convection is increased equivalent to potential temperature advection in combination with thermally-induced slope and valley flows. The importance of surface forcing increases from (1) to (3) but, during summertime, the interaction of all three forcing mechanisms is responsible for the development of specific convergence lines resulting in a complex distribution of precipitation ([2,6,21], and [28]). The results demonstrated that a substantial increase of forecast skill can be achieved if the models are operated at convection-permitting resolution without the need of the parameterization of deep convection. The first downscaling experiment to the convection permitting scale within WRFCLIM on the NEC Nehalem Cluster [26] shows, that regional climate modeling benefits from this resolution, too, as similar systematic errors are present such as the windward-lee-effect in orographic terrain.

2 Simulation with WRF-3.1.0 on the NEC Nehalem Cluster

The results of this climate simulation with WRF Version 3.1.0 are published by Warrach-Sagi et al. [26] and Greve et al. [16]. Here a summary of the model set-up and some results are given. Version 3.1.0 of the WRF model has been applied to Europe on a rotated latitude-longitude grid with a horizontal resolution of 0.11° and with 50 vertical layers up to 20 hPa with the land surface model NOAH [4,5]. The model domain (red frame in Fig. 1) covers the area specified in CORDEX. ERA-interim forcing data is available at approx. 0.75°, and WRF was applied, one-way nested, in a double nesting approach on 0.33° (black frame in Fig. 1) and 0.11°. In an additional experiment for summer 2007, when the Convective and Orographically-induced Precipitation Study (COPS) [28] took place, a third domain with 0.0367° (~4 km) resolution was nested into the 0.11° domain (white frame in Fig. 1), and a convection permitting simulation was performed. The ERA-interim data set is the latest ECMWF global atmospheric multi-decadal reanalysis using a 6-h 4D-Var data assimilation system. Dee et al.[9] give a detailed description and analysis. At the time of the beginning of the CORDEX simulations, ERA-interim data was available from 1989 to 2008. The WRF simulation was carried out from 1989 to

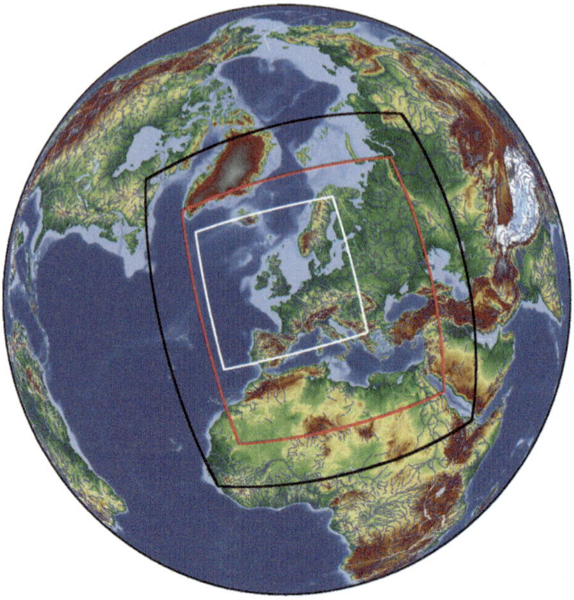

Fig. 1 Domain of WRF for CORDEX Europe on a rotated grid with 0.33° (*black frame*), 0.11° resolution (*red frame*) and 0.0367° (*white frame*)

2008, forced with 6-hourly analysis data at the lateral boundary and daily sea surface temperatures- both from ERA-interim. WRF was compiled at HLRS with PGI 9.04 with openMPI libraries. This is because PGI is twice as fast as gfortran and since ifort compilation is not suggested by WRF staff. At 0.11° the EURO-CORDEX model domain covered 450*450*54 grid boxes (Fig. 1, red box) and the model was run with a time step of 60 s from 1989 to 2008. For the simulation 20 nodes were requested (160 cores), the raw output data was about 80 TB. Within 24 h on the NEC Nehalem Cluster it was possible to simulate 2.5 months. The whole simulation (including waiting time between restarts) took ~8 months and ~800,000 CPU hours.

For the convection permitting simulation experiment a domain of 800*800*54 grid boxes of approx. 4 km horizontal resolution was nested in the EURO-CORDEX model domain (white frame in Fig. 1). A simulation with a model time step of 20 s was run from 15th Mai to 1st September 2007. The simulation took 1 month on the NEC Nehalem Cluster.

3 Simulation with WRF-3.3.1 on the CRAY X6

Precipitation and soil moisture results of this simulation were analysed [16, 26] and due to the results in the beginning of 2012 the simulation was repeated with an updated version of WRF (Version 3.3.1) on the CRAY X6 at HLRS. WRF was

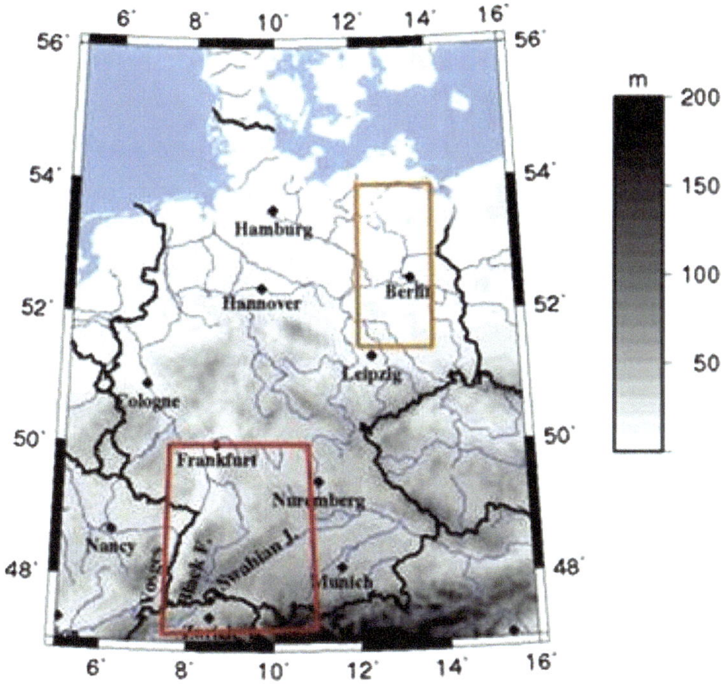

Fig. 2 Map of Germany, the evaluation area from Warrach-Sagi et al. [26] of SW-Germany and NE-Germany are indicated by a *red* and *orange frame*

compiled at HLRS with PGI 11 with openMPI 1.4. This simulation was carried out for the 0.11° domain (red frame in Fig. 1) without nesting to an intermediate grid but a 30 grid box wide boundary. The simulation of the 480*468*54 grid boxes with a 60 s model time step was run from 1987 to 2009 on 40 nodes (1,280 cores), within 24 h 5 months were simulated. Within 3 months the simulation was completed and approx. 90 TB of raw output data were produced. This simulation was completed in April 2012 and is currently under evaluation e.g. within the EURO-CORDEX (www.euro-cordex.net) ensemble of regional climate models [25].

4 Results

For Germany (Fig. 2), the German Weather Service (DWD) processed a consistent 1 km^2 gridded dataset of daily precipitation (REGNIE = Regionalisierung von Niederschlagsdaten) from 1961 to 2009. REGNIE is generated from about 1,200 precipitation measurement stations interpolated on a 1×1 km^2 grid over Germany. During the interpolation, also the station elevation and exposition are considered. Richter [24] gives the climatological (1961–1990) monthly mean undercatch of the

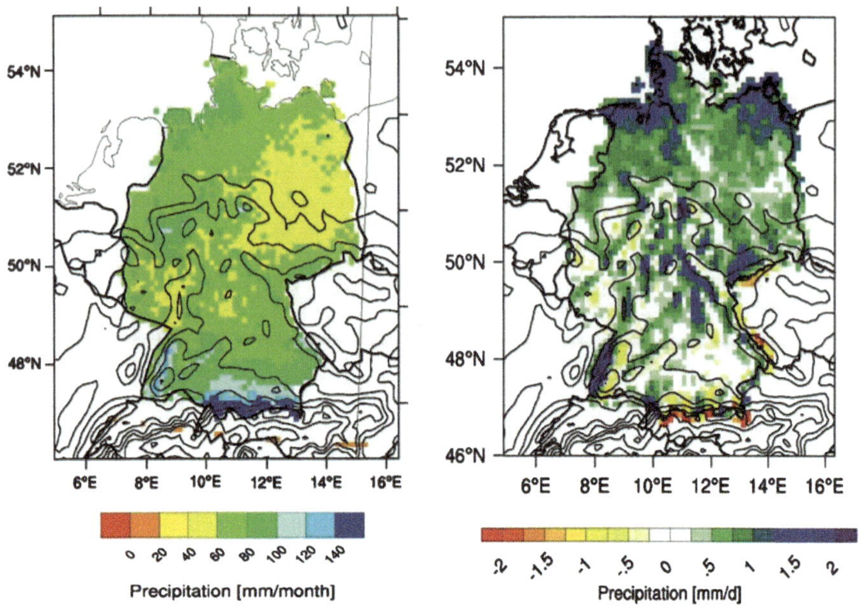

Fig. 3 Mean summer precipitation for 1989–2008 from the REGNIE data (*left*) and the difference of the WRF 3.1.0 simulation to REGNIE on the 0.11° grid (*right*)

german precipitation gauges used in the REGNIE data. It is between 5.6 % in July in very protected locations below 1,000 m in southern Germany up to 33.5 % in February below 700 m at non-protected gauges in eastern Germany. The annual mean precipitation between 1961 and 2009 varied between 584 mm in 1976 and 1,005 mm in 2002. Figure 3 shows the mean summer precipitation for 1989–2008 from the REGNIE data and the difference of the WRF 3.1.0 simulation to REGNIE on the 0.11° grid. A wet bias is seen in the northern half of Germany, especially in the coastal areas. Further the so called windward-lee effect is seen at the mountain ranges, i.e. a wet bias in the windward side and a dry bias on the lee side of the mountain ranges. This is especially obvious in the black forest region in southwest Germany. For hydrologists e.g. it is important that the precipitation is simulated in the correct river catchment. Systematic biases in precipitation amounts may be corrected (e.g. [12]), but it is problematic, if the rain falls in the wrong catchments, and catchments e.g. devide at the top of mountain ranges. From Fig. 4 it can be seen that this windward-lee effect vanishes in a convection permitting simulation, but a dry bias is visible.

For EURO-CORDEX WRF-3.3.1 was applied at 0.11° resolution forced with ERA-Interim data at the lateral boundaries. Figure 5 shows the mean summer (JJA) and winter (DJF) precipitation from 1989 to 2008 for Germany.

In comparison with the WRF-3.1.0 simulation in summer the mean bias over Germany was reduced, but in the south the simulation with WRF-3.3.1 is wetter.

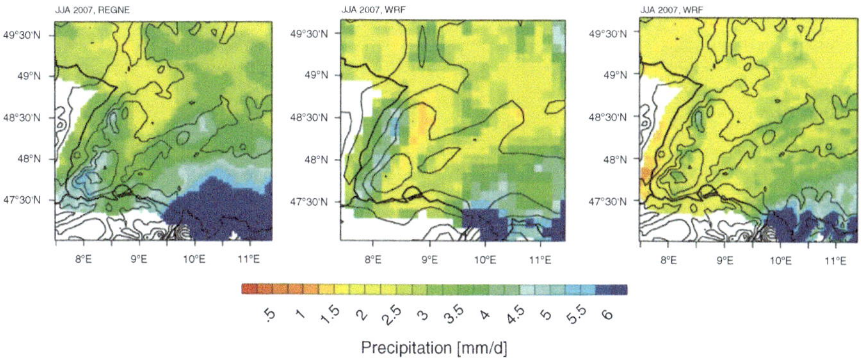

Fig. 4 Summer mean precipitation in 2007 from REGNIE (*left*), WRF-3.1.0 at 0.11° (*middle*) and WRF-3.1.0 at 0.37° (*right*)

In summer the bias is highest, however, in this season a lot of local convection induced precipitation occurs. The lowest precipitation bias is simulated in autumn (not shown). The evaluation of the WRF-3.3.1 simulation is ongoing, mainly in the ensemble evaluation within EURO-CORDEX (e.g. [25] evaluated the simulation of heat waves in Europe).

5 Conclusion

The natural deviation of boundaries from global models from the true state due to model physics and assumed initial conditions, inconsistent physics between global and regional models, and the poor representation of orography and the heterogeneity of land-surface-vegetation properties in RCMs at 10–50 km resolution result in a large gap in our knowledge concerning regional impacts of climate change [11, 17] due to a reduced skill of regional simulations of feedback processes between the land-surface and the atmospheric boundary layer as well as of clouds and precipitation development. Doherty et al. [11], who summarized the research needs that shall follow IPCC AR4 and the new Strategic Plan of the World Weather Research Program (WMO [27]) come to consistent conclusions concerning the promotion of research in two areas: (1) High-resolution, advanced mesoscale atmospheric ensemble modeling, and (2) high-resolution variational data assimilation, e.g., for testing and improving regional climate models in weather forecast mode. The scope of the WRFCLIM project, is to investigate in detail the performance of regional climate projections with WRF in the frame of EURO-CORDEX at 12 km down to simulations at the convection permitting scale within the DFG funded Research Unit 1695. The main objectives of the convection permitting simulations of WRFCLIM are as follows:

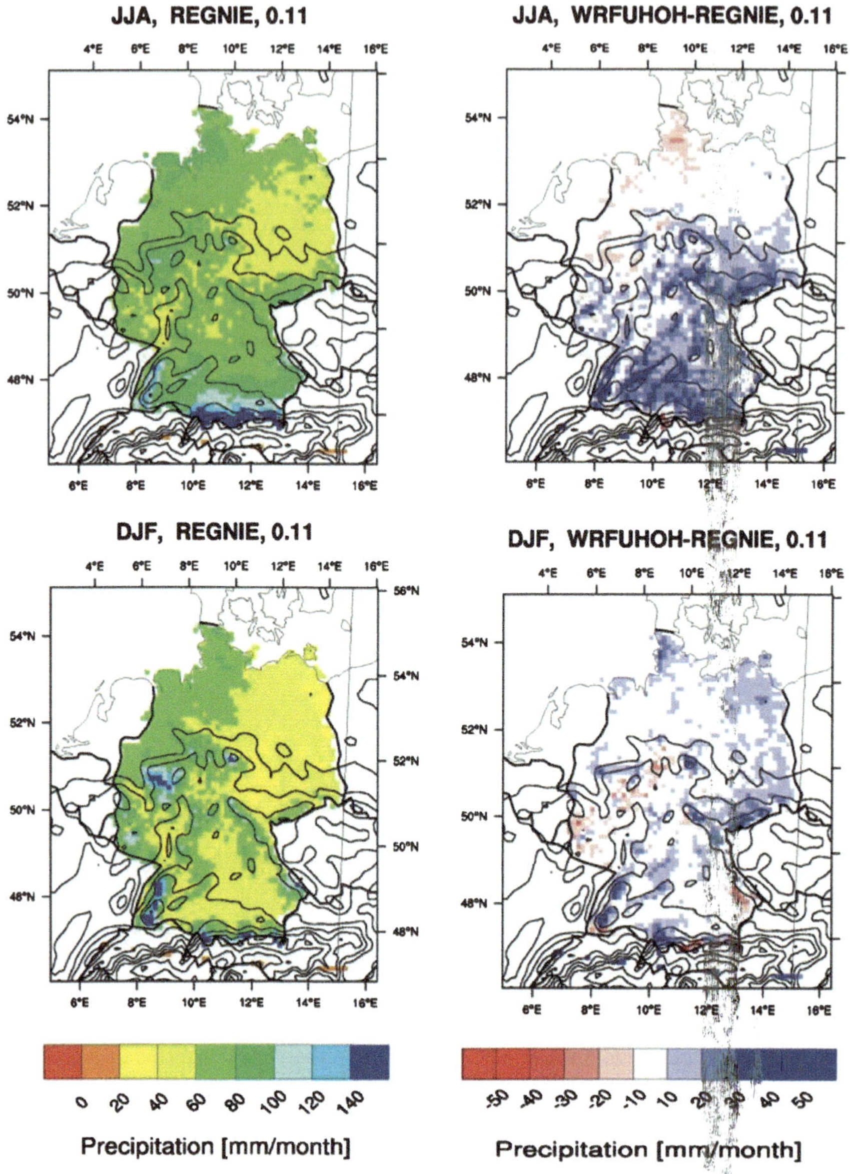

Fig. 5 The mean summer (JJA, *top panels*) and winter (DJF, *bottom panels*) precipitation from 1989 to 2008 for Germany from REGNIE (*left*, interpolated to the model grid) and the difference between the WRF-3.3.1 simulation and REGNIE

- Replace the convection parameterization by the dynamical simulation of the convection chain to better resolve the processes of the specific location and actual weather situation physically.
- Gain of an improved spatial distribution and diurnal cycle of precipitation through better land-surface-atmosphere feedback simulation and a more realistic representation of orography to support the interpretation of the 12 km climate simulations for local applications e.g. in hydrological and agricultural management.

Special attention will be paid to the land-surface-vegetation-atmosphere feedback processes. Further, the model will be validated with high-resolution case studies applying advanced data assimilation to improve the process understanding over a wide range of temporal scales. This will also address whether the model is able to reasonably represent extreme events.

Acknowledgements Kirsten Warrach-Sagi thanks the German Science Foundation for her funding within the frame of the integrated research project PAK 346/FOR 1695 Structure and function of agricultural landscapes under global climate change – Processes and projections on a regional scale. Further we acknowledge the REGNIE data from the German Weather Service. The authors thank the HLRS staff for the permission and support of the simulations on the High Performance Computer in Stuttgart. The simulations were carried out in collaboration with the WESS (Water and Earth System Science) Consortium funded by the BMBF and UFZ Leipzig.

References

1. Beniston, M., D.B. Stephenson, O.B. Christensen, C.A.T. Ferro, C. Frei, S. Goyette, K. Halsnaes, T. Holt, K. Jylhü, B. Koffi, J. Palutikoff, R. Schöll, T. Semmler, and K. Woth, 2007: Future extreme events in European climate; an exploration of Regional Climate Model projections. Climatic Change 81, 71–95.
2. Behrendt, A., S. Pal, F. Aoshima, M. Bender, A. Blyth, U. Corsmeier, J. Cuesta, G. Dick, M. Dorninger, C. Flamant, P. Di Girolamo, T. Gorgas, Y. Huang, N. Kalthoff, S. Khodayar, H. Mannstein, K. Träumner, A. Wieser, and V. Wulfmeyer, 2011: Observation of Convection Initiation Processes with a Suite of State-of-the-Art Research Instruments during COPS IOP8b. COPS Special Issue of the Q. J. R. Meteorol. Soc. 137, 81–100, DOI:10.1002/qj.758.
3. Brockhaus, P., D. Lüthi, and C. Schär, 2008. Aspects of the Diurnal Cycle in a Regional Climate Model. Meteorol. Z., 17 (4), 433–443.
4. Chen, F., and J. Dudhia, 2001a. Coupling an advanced landsurface/ hydrology model with the penn state NCAR MM5 modeling system. Part I: Model implementation and sensitivity. Mon Weather Rev, 129, 569–585.
5. Chen, F. and J. Dudhia,2001b. Coupling an advanced landsurface/ hydrology model with the penn state NCAR MM5 modeling system. Part II: Preliminary model validation. Mon Weather Rev, 129, 587–604.
6. Corsmeier, U., N. Kalthoff, Ch. Barthlott, A. Behrendt, P. Di Girolamo, M. Dorninger, F. Aoshima, J. Handwerker, Ch. Kottmeier, H. Mahlke, St. Mobbs, G. Vaughan, J. Wickert, and V. Wulfmeyer, 2011: Driving processes for deep convection over complex terrain: A multi-scale analysis of observations from COPS-IOP 9c. COPS Special Issue of the Q. J. R. Meteorol. Soc. 137, 137–155, DOI:10.1002/qj.754.

7. Christensen, J.H., and O.B. Christensen, 2007: A summary of the PRUDENCE model projections of changes in European climate by the end of this century. Climatic Change 81, 7–30.
8. Christensen, J.H., T.R. Carter, M. Rummukainen and G. Amanatidis, 2007: Evaluating the performance and utility of regional climate models: the PRUDENCE project. Climatic Change 81, 1–6.
9. Dee, D.P. et al., 2011: The ERA-Interim reanalysis: configuration and performance of the data assimilation system. Q. J. R. Meteorol. Soc. 137, 553–597.
10. Déqué, M., D.P. Rowell, D. Lüthi, F. Giorgi, J.H. Christensen, B. Rockel, D. Jacob, E. Kjellström, M. De Castro, and B. van den Hurk, 2007: An intercomparison of regional climate simulations for Europe: assessing uncertainties in model projections. Climatic Change 81, 53–70.
11. Doherty, S.J., S. Bojinski, A. Henderson-Sellers, K. Noone, D. Goodrich, N.L. Bindoff, J.A. Church, K.A. Hibbard, T.R. Karl, L. Kajfez-Bogataj, A.H. Lynch, D.E. Parker, I.C. Prentice, V. Ramaswamy, R.W. Saunders, M.S. Smith, K. Steffen, T.F. Stocker, P.W. Throne, K.E. Trenberth, M.M. Verstraete, and F.W. Zwiers, 2009: Lessons learned from IPCC AR4: Scientific Developments Needed to Understand, predict and respond to climate change. Bull. Amer. Meteor. Soc., 90, 497–513.
12. Ehret, U., E. Zehe, V.Wulfmeyer, K. Warrach-Sagi, J. Liebert, 2012: HESS Opinions - Should we apply Bias Correction to Global and Regional Climate Model Data? Hydrol. Earth Syst. Sci. Discuss. 9, 5355–5387.
13. Feldmann, H., B. Früh, G. Schädler, H.-J. Panitz, K. Keuler, D. Jacob, and P. Lorenz, 2008. Evaluation of the Precipitation for South-western Germany from High Resolution Simulations with Regional Climate Models. Meteorologische Zeitschrift 17, 455–465.
14. Früh, B., H. Feldmann, H.-J. Panitz, G. Schädler, D. Jacob, P. Lorenz, and K. Keuler, 2010: Determination of precipitation return values in complex terrain and their evaluation. J. Climate 23, 2257–2274. doi: 10.1175/2009JCLI2685.1.
15. Giorgi, F., C. Jones, and G. Asrar, 2009: Addressing climate information needs at the regional level: The CORDEX framework. WMO Bulletin 58, 175–183.
16. Greve, P., K. Warrach-Sagi and V. Wulfmeyer, 2013: Evaluating Soil Water Content in a WRF-NOAH Downscaling Experiment. J. Applied Met. and Climatol., submitted.
17. IPCC, 2007: Climate change 2007: The physical science basis. Solomon, S., D. Qin, M. Manning, L. Chen, M. Marquis, K.B. Avery, M. Tignor, and H.L. Miller (eds.), Cambridge University Press, Cambridge, 996pp.
18. Jacob, D., L. Bähring, O.B. Christensen, J.H. Christensen, S. Hagemann, M. Hirschi, E. Kjellström, G. Lenderink, B. Rockel, C. Schär, S.I. Seneviratne, S. Somot, A. van Ulden, and B. van den Hurk, 2007: An intercomparison of regional climate models for Europe: Design of the experiments and model performance. PRUDENCE special issue, Climatic Change, 81, Supplement 1, May 2007.
19. Jaeger, E.B., I. Anders, D. Lüthi, B. Rockel, C. Schär, and S. I. Seneviratne, 2008: Analysis of ERA40-driven CLM simulations for Europe. Meteorol. Z. 17, 349–367.
20. Morrison, H. and A. Gettelman, 2008: A new two-moment bulk stratiform cloud microphysics scheme in the Community Atmosphere Model, version 3 (CAM3). Part I: Description and numerical tests. J. Climate, 21, 3642–3659.
21. Schwitalla, T., H.-S. Bauer, V. Wulfmeyer, and F. Aoshima, 2011: High-resolution simulation over central Europe: Assimilation experiments with WRF 3DVAR during COPS IOP9c. Q. J. R. Meteorol. Soc. 137, 156–175, DOI:10.1002/qj.721.
22. Skamarock, W.C., J.B. Klemp, J. Dudhia, D.O. Gill, D.M. Barker, M.G. Duda, X.-Y. Huang, W. Wang, and J.G. Powers, 2008: A description of the Advanced Research WRF version 3. NCAR Tech Note, TN-475+STR, 113pp.
23. Thuburn, T.: Some conservation issues for dynamical cores of NWP and climate models. J. Comp. Phys., 227 (2008), 3715–3730.
24. Richter D. (1995) Ergebnisse methodischer Untersuchungen zur Korrektur des systematischen Messfehlers des Hellmann-Nie-derschlagsmessers. Berichte des Deutschen Wetterdienstes 194: 93 pp

25. Vautard, R., A. Gobiet, D. Jacob, M. Belda, A. Colette, M. Deque, J. Fernandez, M. Garcia-Diez, K. Goergen, I. Guettler, T. Halenka, K. Keuler, S. Kotlarski, G. Nikulin, M. Patarcic, M. Suklitsch, C. Teichmann, K. Warrach-Sagi, V. Wulfmeyer and P. Yiou, 2013: The simulation of European heat waves from an ensemble of regional climate models within the EURO-CORDEX project, Climate Dynamics, 10.1007/s00382-013-1714-z.
26. Warrach-Sagi, K., T. Schwitalla, V. Wulfmeyer and H.-S. Bauer, 2013: Evaluation of a CORDEX-Europe simulation with WRF: precipitation in Germany, Climate Dynamics, DOI 10.1007/s00382-013-1727-7
27. WMO, 2010: Strategic plan for implementation of WMO's World Weather Research Programm 2009–2017. World Meteorological Organization, Geneva, Switzerland. Available online at: http://www.wmo.int/pages/prog/arep/wwrp/new/documents/final_WWRP_SP_6_Oct.pdf.
28. Wulfmeyer, V., A. Behrendt, Ch. Kottmeier, U. Corsmeier, C. Barthlott, G.C. Craig, M. Hagen, D. Althausen, F. Aoshima, M. Arpagaus, H.-S. Bauer, L. Bennett, A. Blyth, C. Brandau, C. Champollion, S. Crewell, G. Dick, P. Di Girolamo, M. Dorninger, Y. Dufournet, R. Eigenmann, R. Engelmann, C. Flamant, T. Foken, T. Gorgas, M. Grzeschik, J. Handwerker, C. Hauck, H. Höller, W. Junkermann, N. Kalthoff, C. Kiemle, S. Klink, M. König, L. Krauss, C.N. Long, F. Madonna, S. Mobbs, B. Neininger, S. Pal, G. Peters, G. Pigeon, E. Richard, M.W. Rotach, H. Russchenberg, T. Schwitalla, V. Smith, R. Steinacker, J. Trentmann, D.D. Turner, J. van Baelen, S. Vogt, H. Volkert, T. Weckwerth, H. Wernli, A. Wieser, and M. Wirth, 2011: The Convective and Orographically Induced Precipitation Study (COPS): The Scientific Strategy, the Field Phase, and First Highlights. COPS Special Issue of the Q. J. R. Meteorol. Soc. 137, 3–30, DOI:10.1002/qj.752.

GPSR Compliance

The European Union's (EU) General Product Safety Regulation (GPSR) is a set of rules that requires consumer products to be safe and our obligations to ensure this.

If you have any concerns about our products, you can contact us on ProductSafety@springernature.com

In case Publisher is established outside the EU, the EU authorized representative is:

Springer Nature Customer Service Center GmbH
Europaplatz 3
69115 Heidelberg, Germany

Batch number: 10584666

Printed by Printforce, the Netherlands